# Water Supply Planning

# Water Supply Planning

David W. Prasifka

Krieger Publishing Company
Malabar, Florida
1994

Original Edition 1994
(Based on *Current Trends in Water-Supply Planning*)

Printed and Published by
**KRIEGER PUBLISHING COMPANY**
**KRIEGER DRIVE**
**MALABAR, FLORIDA 32950**

**Library of Congress Cataloging-In-Publication Data**

Prasifka, David W., 1951–
 Water supply planning/David W. Prasifka.
  p.  cm.
 Based on Current trends in water-supply planning.
 Includes bibliographical references and index.
 ISBN 0-89464-838-1 (acid-free paper)
 1. Water—supply—Planning.  I. Title.
TD353.P7  1994
333.91'15—dc20

93-28392
CIP

10 9 8 7 6 5 4 3 2

I have spoken of the rich years when the rainfall was plentiful. But there were dry years too, and they put a terror on the valley. The water came in a thirty-year cycle. There would be five or six wet and wonderful years when there might be nineteen to twenty-five inches of rain, and the land would shout with grass. Then would come six or seven pretty good years of twelve to sixteen inches of rain. And then the dry years would come, and sometimes there would be only seven or eight inches of rain. The land dried up and the grass headed out miserably a few inches high and great bare scabby places appeared in the valley. The live oaks got a crusty look and the sagebrush was gray. The land cracked and the springs dried up and the cattle listlessly nibbled dry twigs. Then the farmers and the ranchers would be filled with disgust for the Salinas Valley. The cows would grow thin and sometimes starve to death. People would have to haul water in barrels to their farms just for drinking. Some families would sell out for nearly nothing and move away. And it never failed that during the dry years the people forgot about the rich years, and during the wet years they lost all memory of the dry years. It was always that way.

JOHN STEINBECK
*East of Eden*

# Contents

PREFACE TO THE 1994 EDITION                                    ix

PREFACE                                                        xi

INTRODUCTION                                                    1
*The World of Water*

1. WATER-DEMAND CHARACTERISTICS                               19
   *Water-Demand Criteria • Water-Demand Components*

2. WATER-DEMAND FORECASTING                                   72
   *Data Collection • Forecast Methods • Forecast Techniques*
   *• Impacts of Water Conservation • Impacts of Water Quality*
   *• Integrated Approach to Demand Forecasting*

3. WATER PRICING                                             153
   *Pricing Policies • Water Rates and Price Structures • Price*
   *Elasticity of Demand*

4. WATER-LIFELINE HAZARD MITIGATION                          192
   *Water-Lifeline Hazards • Economic Impact of Hazard Mitigation*
   *• Hazards and Public Policy • Seismic Hazard Mitigation: Case*
   *Studies • Drought Management*

5. SOCIAL OBJECTIVES IN WATER RESOURCES
   PLANNING AND MANAGEMENT                                   239
   *Preferred Planning Approach • Social and Environmental*

*Objectives • The Next Twenty Years • Technology-Based Industry • Long-Term Supplemental Water Supplies*

BIBLIOGRAPHY                                              253

INDEX                                                     257

# Preface

Since the publication of my previous book in 1988, the water supply situation in the world has steadily deteriorated. Drought of unheard of proportions has plagued less-developed and industrial countries around the globe. Famine has struck many parts of Africa. Saline water intrusion is threatening coastal population centers in the Philippines. Domestic water reservoirs in China are becoming so putrified that they are being converted to wastewater storage. Potable water in Poland is so polluted that mothers boil the water before bathing their infants. In southern Iran, groundwater supplies are so polluted that in some areas no crop has been harvested in the last three years. The world's largest freshwater lake, Lake Baikal in the Russian border region near Mongolia, has been contaminated by an estimated two billion cubic yards of industrial and municipal waste since 1966. Outbreaks of blue-green algae in Australia threaten the domestic water supply and are called that county's worst ecological disaster. The great cathedral in Mexico City is sinking because of overdrawn groundwater.

All of these events have taken place in an era of instant global communication where the problems of one continent are known immediately to the rest of the world. The 1990s are seeing a global rise in consumerism and environmentalism. Consumers everywhere are demanding the same level of water quality and environmental protection whether they live in Chicago, Cannes, Canturbury, or Calcutta.

The fall of communism has also opened new frontiers to the technology push. Communism's trashing of the environment has left a legacy of pollution and contamination which threatens the global water supply and creates a huge demand for the advanced technologies of the western countries.

Accordingly, I have expanded the focus in this book to include the global water situation. The introductory section "The World of Water" looks at water supply opportunities and constraints around the globe. The next four chapters address key issues and concepts in water supply planning, such as: regionalization, water supply planning philosophies, water demand forecasting methods and techniques; water rates and price structures; hazard mitigation; and social and environmental impacts. The final chapter concludes with a discussion of water supply planning during the next twenty years. Once again, a global focus is used to explore innovative and creative approaches to water management around the world.

Finally, a series of excellent publications by the United Nations provided an invaluable insight into water supply problems and opportunities around the world.

David Prasifka
Irvine, California USA

# Preface

Mounting evidence suggests that the nation's water-supply systems are under considerable stress. Surface-water supplies have become inadequate or unreliable. Groundwater supplies have been exhausted or contaminated. Rising energy costs and fluctuations in the cost of money have resulted in substantial increases in water rates since the early 1970s. Environmental constraints and an increased concern for water quality have led to new regulations and more stringent water quality standards. Most important, more people are involved in making decisions about our water supply.

Many of the current problems with water supply systems have evolved from the historical development of water utilities. Many water utilities were established in areas that had inexpensive water supplies, in anticipation of rapid population and industrial growth. The early charges for water were deliberately set below the actual cost of supply in order to encourage water use, attract industry, and improve public health and fire-fighting capabilities. Water systems, except for those serving the largest cities, remained small in comparison to other utilities. Most water systems were not interconnected because of the high cost involved in laying pipelines across sparsely populated areas.

During the 1970s, the cost of supplying water rose dramatically because of increased energy costs for pumping, increased prices of chemicals essential for treatment, high inflation rates, and increased maintenance costs for aging infrastructures. The demands on what had already become an overextended water-supply system were intensified by population growth and industrial development, as well as by the loss of supplies through contamination from hazardous wastes and other sources. The increased cost of providing water was only partially met by water rate increases; larger subsidies from tax revenues covered the remainder of the costs. Because of aging systems and inadequate revenues to cover costs, many water utilities deferred necessary maintenance. Consequently, fixed assets—especially water mains—deteriorated.

In the 1980s, strict limitations on property taxes, together with federal cutbacks, have further reduced the funds available to municipalities. Unlike other utilities, water utilities have little opportunity to reduce costs through technological change, because most water-system facilities have long lives. This has resulted in a substantial increase in investment as plants built many years ago at relatively low construction costs are replaced at today's higher costs and interest rates. In addition to replacement costs, new facilities built to increase system capacity usually cost more per unit of output because new sources of supply are farther away, more costly to develop, and (probably) of poorer quality.

The purpose of this book is to examine current trends in water-supply planning that have evolved in response to the changing political and economic climate of the 1980s. These trends are embodied in myriad issues and concepts that have emerged as a result of and in response to the dwindling of developable freshwater supplies in the nation. By incorporating these issues and concepts into the planning process, water managers can lower the level of risk associated with the development, protection, and use of water.

The primary issue faced by water managers in the 1980s is the need for reliable forecasts of water needs. In the past, mistakes (or at least overestimates) in planning were easily covered up, because, with rapid growth, the excess capacity would eventually be needed anyway. Today, however, inaccurate forecasts may lead to severe economic and environmental costs. A reliable forecast must consider the impacts of water conservation, water reclamation, unaccounted-for water uses, the price elasticity of water demand, and other concepts previously dismissed by water managers.

Other issues covered in this book include: regionalization; the selection of a water-supply planning philosophy; a comparative evaluation of forecasting methods and techniques; a discussion of water rates and price structures; hazard mitigation and its impact on public policy; public participation in the planning process; and the determination of the social acceptability of a proposed water-supply project or program. The emphasis is on identifying the practical applications of these issues and on gaining an understanding of the concepts that are used to describe them.

The material presented in this book will be of interest and benefit to engineers, planners, water superintendents, government and regulatory officials, water plant operators, utility managers, administrators, and others faced with the daily task of furnishing or ensuring supplies of high-quality water to the public. The book will also be valuable as a supplemental reference text for teachers and students who are interested in studying the practical applications of water-resources engineering in planning for the ultimate water needs of a community.

The most important single source of material for this book came from the U.S. Army Corps of Engineers—primarily the Institute for Water Resources (IWR) in Ft. Belvoir, Virginia. The IWR offers a number of up-to-date publications to assist water utilities and engineers with planning, design, and operational problems. The reader is directed to the bibliography following the last chapter for a comprehensive list of IWR publications used in this book.

I would like to thank the following organizations for allowing me to use material from their publications: American Water Works Association; American Society of Civil Engineers; Virginia Water Resources Research Center; Institute of Governmental Studies at the University of California, Berkeley; U.S. Water News; Freshwater Foundation; and American Geophysical Union.

Finally, I would like to thank my wife, Carol, for her untiring support throughout the three years of research that went into this publication.

# Water Supply Planning

## INTRODUCTION

# The World of Water

An intergalactic traveler who happened to stumble upon our solar system, whether by accident or by choice, would make several startling observations. His eyes would immediately be drawn to the third planet revolving around a glowing star. In his travel journal, he would refer to this planet as the blue planet. His trained eye would tell him that this characteristic blue color, which distinguishes it from the other planets in the solar system, results from vast quantities of water stored in the oceans which cover a majority of the planet's surface.

He would excitedly make the appropriate notations in his journal, marking the precise location of this watery oasis among the bleak expanses of space, and proceed to his destination. In another time, measured by means unknown today, he would return to this outpost in space. He would know that the treasure he had discovered there would be waiting for him in the same quantity he observed that day. He would know this for a fact because, unlike other natural resources discovered in the chartered territories of stars and solar systems, water is a renewable resource.

Like energy, water can neither be increased (like fish, timber, etc.), nor diminished (like coal, gas, etc.). Water can only be transferred from one state to another. Indeed, water is continuously converted from one state to another (solid-liquid-vapor) through nature's hydrologic cycle.

Here on Earth, as we await the return of this intergalactic traveler, much has happened which threatens our most valuable resource.

The population of our planet is now 27 times greater than it was at the time of the birth of Christ. Each day, almost 400,000 people are born. Only 200 million people were on Earth at the time of the birth of Christ. By the time Europeans first settled in America, 1,600 years later, world population had grown to 500 million. In the next 250 years, by 1850, human population grew to 1 billion, and then more than doubled in the next 100 years to 2.5 billion. Now, in just the last 40 years, Earth's population has more than doubled again to 5.4 billion people.

The United Nations now projects that if fertility ultimately stabilizes at a replacement rate of about 2.06 births per woman, the global population will reach 10 billion in the year 2050. But those numbers could vary greatly if fertility rates actually turn out to be higher or lower. At a rate of 2.5 births per woman, the U. N. calculates world population would reach 28 billion in 2150.

World population grew by a record 92 million in 1991, a result of 143 million births and 51 million deaths, according to the Bureau of Census. Indian and China led the way with 17 million and 16 million, respectively. However, the fastest growing regions of the globe are Africa and the Middle East, each with its own set of problems. Africa has the fastest growing population of any continent in history, while at the same time its per capita grain production has fallen nearly 20 percent. Severe water shortages are emerging in the Middle East which could rival oil as a basis for war.

Global municipal water use is estimated to be 35 times today what it was three centuries ago. People withdraw the equivalent of Lake Huron (with a storage volume of 3,580 cubic kilometers ($km^3$) from the world's rivers, streams, lakes, and aquifers each year, and withdrawals have been increasing by 4 to 8 percent a year in recent decades. About 40 percent of the water withdrawn is returned to the water cycle as wastewater. There are currently two billion people in the world who don't have an adequate water supply. Three billion people are without adequate sanitation.

Our freshwater resources are under increasing stress. According to the World Resource Institute, about two-thirds of global withdrawals are used for agriculture and about one-fourth for industry. By the end of the century, estimates are that withdrawals for agriculture will increase only slightly, while industrial withdrawals will likely double. Industrial development and population growth will also add pollutants to fresh water, unless governments boost their efforts to treat wastewater or prevent pollution.

Each country, to supply its population and civilization, needs about 2,744 liters (725 gallons) a day of replenishable water supply per person. The Earth holds about 10 times the amount of water required to meet this need; however, it is not distributed evenly over populated areas. Worldwide water use has tripled since the 1950s, and twenty-six countries now have more people than their water supplies can adequately support. Eleven African nations are among these twenty-six nations that now fall below the per-person level of 2,744 liters per day. South Africa, Sudan, Morocco, and Malawi will join the list by the end of the century.

The Worldwatch Institute reports that from the first irrigation practices several thousand years ago in the Middle East, irrigation grew to some 93 million hectares (230 million acres) worldwide by 1950. In 1978, when the irrigated area per person peaked at about 49 hectares (120 acres) per thousand people, the amount of irrigated land more than doubled to 206 million hectares (509 million acres). Since then, until 1989, worldwide irrigation growth slowed dramatically, reaching 235 million hectares (580 million acres), a gain of 14 percent in eleven years.

Because global population is increasing at such a rapid pace, the irrigated acreage per person has reduced 6 percent since 1989. The Institute estimates it unlikely that rapid growth in global irrigation will be reestablished. If any gains are made in irrigation, they likely will be made through improved irrigation efficiency.

Let us look at the world of water and examine the problems and opportunities facing water supply development.

Water covers about three-quarters of the earth's total surface area. However, the disproportionate distribution of water, especially of freshwater, is striking. For example, the Amazon river system carries one-fifth of all waters discharged by the world's river systems. Water carried by the Amazon river system alone is theoretically adequate to satisfy the world's demands twice over. China's mighty river—the Yangtze Kiang—carries enough water per year to provide every person on this planet with about 500 litres (132 gallons) of water every day. Yet, in some 26 percent of the earth's total land surface, the potential evaporation far exceeds the average precipitation; some 18 percent of total land areas are classified as semi- or complete desert. The Amazon River originates from the Andes, yet the other side of this formidable mountain range constitutes some of the driest places on earth.

Water plays a very significant role in the socioeconomic development of a country. The efficient, effective, and rational development of water resources is crucial to a country's ability to meet its growing demand for food and energy. Consider the following examples:

## INDIA

India has 14 major rivers, each with catchment areas larger than 20,000 square kilometers (sq. km.). In addition, there are some 110 minor rivers and innumerable small streams and creeks in the country. India also has some 148 notable lakes.

The availability of water is highly uneven, both in space and time. Out of a total estimated precipitation of 400 million hectare meters in the country, the surface water availability is about 178 million hectare meters. Out of this, only 50 percent can be put to beneficial use because of topographical and other constraints. In addition, there is a groundwater potential of about 42 million hectare meters. Precipitation is confined to the monsoon season which is three to four months in a year and varies from 10 centimeters (cm) to over 1,000 cm.

Floods and drought affect vast areas of the country. A third of the country is drought-prone, while floods affect an average of around 9 million hectares per year. According to the National Commission on Floods, the area susceptible to inundation is about 40 million hectares. Due to this uneven distribution of water in respect to space and time, shortages of drinking water are experienced in many parts of the country.

In India, about 75 percent of the population resides in 583,000 scattered villages where drinking water quality monitoring in almost non-existent. In urban areas, which account for 25 percent of the population, the water supply undergoes conventional treatment including disinfection. However, a large proportion of India's wastewater, whether domestic or industrial, treated or untreated, ultimately manages to get into the country's river systems. The nature of the impact of wastewater on a river depends upon its characteristics and pattern of travel. Although the sea is the ultimate sink, it is the river that plays a major role in transport.

In many parts of the country, the available water is unfit for drinking, mainly for two reasons:

1. Biological contamination resulting in guinea worm disease and other waterborne diseases like cholera, typhoid, diarrhea, infectious hepatitis, bacillary and amoebic dysentery, and gastroenteritis.
2. Chemical contamination resulting in unacceptable concentrations of total dissolved solids, fluorides and iron.

In order to resolve these problems, the government of India has undertaken a program of establishing water quality testing laboratories. The program envisages an integrated approach towards water, health and sanitation, water quality surveillance, and monitoring and training of personnel.

Foreign assistance is envisaged from the World Bank, UNICEF, UNDP, the Asian Development Bank, plus bilateral assistance from countries such as Sweden, the United Kingdom, Denmark, etc. Assistance will focus on integrated demonstration projects involving water supply, sanitation, health education, social communication, etc. A key feature of all projects will be to allow international agencies to act as catalysts to promote community participation and awareness and to introduce new technology that is both appropriate and cost-effective.


## AFGHANISTAN

Afghanistan is a semi-arid, land-locked country situated in the south central part of the Asian mainland. The surface area of the country is about 65.6 million hectares. Only 12 percent of the total land is arable. Of this, only 50 percent is cultivated in a normal year. Forests cover 1.9 million hectares and pastures another 30 million hectares. Only about one-eighth of the geographical area of the country is cultivatable. Land is therefore scarce, but scarcer still is water.

The population is estimated at about 16.7 million, of which 2.7 is urban, 12.6 million rural and 1.5 million nomadic. The rural population is scattered over 30,000 villages. The density of the population is higher in the cities, especially in Kabul, the capital of the country which has 1.4 million people.

Water is the key input for agriculture which is the backbone of the country's economy. Directly or indirectly, about 80 percent of the population derives its livelihood from agriculture and livestock, handicrafts, etc., which contribute about 58 percent to the Gross National Product (GNP) of the country and accounts for over 50 percent of export earnings. An increase in agricultural production is therefore the main hope for an improvement in living conditions for a large section of the poor people located in the rural areas.

Afghanistan's major water source is from the rivers fed by snow melting in the central mountains. Due to the lack of water storage facilities, plus poor water management practices, most of the river water goes to waste. Only 10–15 billion cubic meters out of about 50–55 billion cubic meters is being used at present annually. With arid conditions, unirrigated cultivation is almost impossible, although dry farming is sometimes attempted with low and uncertain yields.

Safe drinking water is a scarce commodity in the country. Only about 2.5 million people (15 percent) out of a total population of 16.7 million receive safe drinking water which includes 900,000 urban and 1.6 million rural population. These figures are very low, even compared with other least developed countries. There is no central sewage system anywhere.

To provide safe drinking water for the public, to protect the country's water resources and to have control over water quality, assistance is needed from international institutions and countries who are more advanced in technical matters.

## ISLAMIC REPUBLIC OF IRAN

Most natural waters in Iran flow from the humid mountains in the north to the arid land in the south. Precipitation, when it comes, is often of a very intensive nature and short duration. Statistical records show an average duration of not more than 15 minutes. The same records also show that the mean precipitation in the mountains in the north is 1,500 millimeters (mm), while the southern part of the country receives as little as 100 to 150 mm per year. Of the total precipitation, about 60 to 70 percent evaporates; 10 to 15 percent constitutes surface water runoff and the rest infiltrates as groundwater.

Both surface water and groundwater have become severely polluted by activities such as human settlement, industrial development and agricultural production. The country lacks treatment plants for both municipal and industrial wastewater; therefore, the level of pollution is very severe. Both surface water and groundwater supplies are seriously affected. The latter, at least in the south, depends on the porosity of soils and relatively shallow groundwater tables, commonly not deeper than 5 meters. The pollution in the south is so severe that in some areas no crop has been harvested in the last three years due to lack of groundwater of the proper quality for irrigation.

The government is aware that measures must be taken as soon as possible to avert a very serious situation. However, Iran has been at war for the last nine years, and thus must give priority to reconstruction of towns, villages and infrastructure, as well as industry and the economy.

## THE PHILIPPINES

The Philippines is a country consisting of over 7,000 islands, with a population of some 65 million. It is rich in natural resources and is blessed with very fertile lands. It has three major island groups—Luzon, Visayas and Mindanao. The Philippines has about 776 rivers of which 248 are classified with regard to their prime or best usage ranging from drinking water supply, contact recreation, fishing, irrigation, and navigation.

The country's drinking water supply is essentially provided by two systems: (1) a centralized surface water system managed by the local waterworks districts, and (2) a decentralized groundwater system consisting of deep wells constructed by the developer of private subdivisions, industries and individual homeowners. The inadequacy of surface water supply to meet the population's water demand, coupled with the failure of the National Water Resource Council to regulate the extraction of groundwater, has resulted in the over-exploitation of the groundwater resources. This has created a serious problem regarding saline water intrusion in some coastal cities.

In the metropolitan Manila area, inland liquid waste from industries, houses, buildings, or waste resulting from human activities finds its way into five rivers that cut across the metropolitan area and eventually discharge into the sea. All five rivers contain low DO (dissolved oxygen concentrations) and high BOD (biochemical oxygen demand), making them suitable only for navigational purposes. In metro Manila, 70 to

80 percent of water pollution is caused by the general public and only about 12 to 15 percent of the population is served by a sanitary sewage system, while the rest depend on improperly maintained septic tanks.

Metro Manila rivers discharge into Manila Bay, and because the mean sea level in Manila Bay is getting higher and higher every year relative to the elevation of the Manila area, these rivers do not actually empty all their pollution loads into the bay. It is only during low tides that some of these pollutants are flushed out to the bay to be diluted or diffused. In this situation, pollutants are just going back and forth along the length of the river with very little chance of escaping into the bay. Because of the now estuarine nature of this river, tides can reach as far as 4 km upstream from the mouth.

Periodic floods during the wet season push the pollutants out of the river and into the bay. However, this just transfers the pollutants to Manila Bay. Therefore, it is probable that the bay, where the tidal current is much slower than on the open sea, will become heavily polluted in the near future. As a matter of fact, beaches around the bay area are no longer safe for bathing. In 1985, the bacterial count on the beaches along the coast and Manila Bay ranged from three to over 100 times greater than allowable standards for recreational waters.

In addition to pollution caused by human activities, the heavy shipping in the Bay is anticipated to carry a high risk of accidental oil spills. During the period August 1975 to December 1976 alone, there were four recorded oil spill accidents in the Bay.

As of 1986, there were 18 rivers throughout the Philippines which were categorized as dead. A dead river is characterized as black in color, anaerobic, heavily silted, and its bank vanished; DO is zero throughout and BOD is greater than 100.

In the Philippines, the Department of Environment and Natural Resources (DENR) is the primary agency responsible for the conservation, management, development and proper use of the country's environment and natural resources.

## VIET NAM

Viet Nam is among those countries that have relatively large surface water bodies and sea coasts in relation to the mainland area. The total water course length is about 40,000 km with roughly 900 million cubic meters annual discharge into the sea. Viet Nam covers a total area of 331,689 square kilometers, and lies within the tropic zone of the northern hemisphere governed by the Asian monsoon system, with abundant rainfall and runoff. The surface runoff is the prime source of water for domestic and industrial uses, irrigation and other applications.

The river network of the country is dense; the two largest rivers are the Mekong River in the south and the Hong River in the north.

Basically, the hydrological conditions of river basins across the country are the same: runoff and striking contrast between dry and wet seasons. The wet season starts in May and ends in October. Its total runoff accounts for 80 percent of the yearly rainfall. The flood season starts a bit later and its runoff percentage is the same as that of rainfall.

The surface water resources in Viet Nam mostly are of good quality for domestic, agricultural, industrial and other uses; however, two natural water quality problems exist. The first is the saline water in the coastal regions, and the second, the acid water in the flooding zones of the Mekong Delta.

High tidal fluctuations and small runoffs during the dry season are favorable conditions for saline water to intrude inland, especially in the Hong and Mekong deltas where the river slopes are small and tidal fluctuations are high.

The total area of all coastal regions covered by saline water during the dry season is over 2 million hectares. In these zones, there is not only a shortage of water for irrigation but the quality of drinking water is also a big problem. These waters are suitable for virtually no uses at all. People there have to store rainfall in tanks and collect fresh water from 30 to 40 kilometers away.

Acid water, mainly in the Mekong Delta, is also a problem in the development of agriculture and fisheries, and in providing water for domestic use. The acid water is related to the pedomorphological conditions of the area. The total area of acid and potential acid soil in the Mekong Delta is about 1.6 million hectares. Also, the acid water in acid zones spreads over areas where soils are of good quality, which influences rice yields and soil properties.

The population density in the Mekong Delta is high ($>300$ people/ $km^2$) and settlements are concentrated along roads, banks of rivers and canals. In the Delta, as well as other areas, there are no wastewater treatment plants in any cities or towns. Domestic wastes are discharged directly into water bodies which, at the same time, are used for washing clothes and kitchenware, bathing and as a source of drinking water. Although simple treatment methods (filtration techniques) are used in the individual households to purify the water for human consumption, the hygiene situation has to be regarded as dangerous to human health.

Concentrations of toxic elements from factories and cities in river reaches near the outlets are very high, far beyond allowable standards, since most waters have not yet been treated.

The Viet Nam government realizes that the key to the reduction or elimination of these water supply constraints is water management. However, appropriate water legislation is still in the process of being developed. For example, international water regulations are currently being studied in order to establish criteria for water use among riparian countries of the Mekong Delta.

There are no existing organizations in Viet Nam that could naturally take on the responsibility of monitoring the environment and enforcing new legislation. Possible structures are local and regional organizations for health, agriculture, fisheries, forestry, and planning.

Water supply problems are not limited to developing countries. The significant changes that have occurred in the global political environment have impacted water management practices in many of the world's industrialized countries. Consider the following examples:

## GERMANY

Germany Unification of Germany in 1989 has had an enormous effect on normalizing water supply and wastewater treatment. Nevertheless, there remain substantial differences.

In the former East Germany, for instance, little investment was made in the distribution system and treatment works for the past few decades. Therefore, a lot of effort is required to remodel and repair those systems to bring them up to West German standards. In the former West Germany, nearly all public water systems are owned by local municipal governments. This presents another contrast to the East. Because of the tremendous capital needs in the eastern parts of the country, some of the water systems may be privatized.

Traditionally, water in Germany is naturally clean. A large number of water utilities in Germany do not disinfect including Munich and Berlin. This is possible because groundwater is the primary source of supply in Germany, and therefore, the water has low regrowth potential and a low assimilable organic carbon.

The key water issue in Germany is quality, not quantity, because the water demand has decreased dramatically in recent years as a result of conservation programs, pricing policies, and wastewater charges that are two to five times higher than for drinking water. Lastly, the environmental movement is encouraging German citizens to save water and energy and has produced a general mood to conserve all natural resources.

Water is expensive in Germany; the price often exceeds $2 per cubic meter and averages $1.50 per cubic meter. This price will likely increase because all political parties agree that water should not be wasted but used with maximum care.

Formerly, the German water utilities led the world for decades in using technologies such as granular activated carbon and biological activated carbon in water treatment. But now, the pendulum is swinging back toward more natural treatment, including biological processes, underground passage, and riverbank filtration. Thus, Germany is concentrating on protecting the raw-water sources and relying less on treatment technology to correct an intrusion on water quality.

## FRANCE

Unlike most other European countries, France has a management system that for more than a century has given local communities the power to choose between direct management and delegated management. Sixty percent of the nation's communities have opted to delegate their water management, primarily to two private companies, Compagnie Generale des Eaux and Lyonnaise des Eaux-Dumez. The distribution of water by these two companies accounts for more than 70 percent of the volume of water distributed in the country.

Over time, these two companies have developed their utility activities to include potable water treatment and supply, wastewater collection and treatment, heat distribution, and industrial and domestic waste collection and treatment for local communities. This practice allows the companies to integrate all aspects of the water use cycle, taking into account growing consumer demands, an increasingly strict regulatory framework, and the need to protect the environment.

In France, drinking water must comply with the directives defined by the Ministry of Health in accordance with the standards laid down by the European Community. These standards are developing towards ever lower tolerance thresholds, which are making tap water a product of even higher quality.

Still, the more than 10,000 water systems in France involve a wide variety of treatment practices, ranging from simple disinfection to the most sophisticated treatment processes. The quality problems that must be tackled essentially come from the concentration of human activities within limited geographical areas.

Private French water distribution companies have become multinational holdings, integrating water issues in industrialized as well as developing countries. These different, specific characteristics of the French water industry provide an extraordinary magnitude to one of the driving forces of evolution—the technology push. Such practices have stimulated competition between the large utilities of each European country.

## POLAND

"Privatization" takes on a different meaning in Poland. In this country, utilities are seeing a steady decentralization in the wake of communism's fall. Privatization refers not to private sector involvement in utility operations, but to municipalities and cities assuming control of their respective portions of the distribution system. The smaller governmental entities are struggling to establish formal procedures for testing, metering, and billing water. Large, regional water systems are being whittled down into smaller, city-based systems.

Polish water utilities are about two decades behind their U.S. counterparts in treatment strategies and management techniques. For example, water in the Vistula River, Warsaw's water supply, is heavily laden with human and industrial wastes and serves as a stark reminder of communism's trashing of the environment. Over the years, the communist Polish government's central plan called for concentrating steel, chemical, and pulp plants along the river. Vistula River water has been known to cause skin irritation and is boiled before drinking. The water's tendency to irritate human skin is due to a high concentration of chlorine necessary to neutralize industrial wastes. Mothers are advised to bathe infants only in tap water that has been boiled for a full 15 minutes.

Outside financial help is a dominant theme for Polish officials. The country simply cannot bankroll the improvements it deems necessary. But lack of coordinated planning, framed against the background of major economic and political change in Poland, are major stumbling blocks in water utilities' path toward change. International investors are reluctant to support a project when the utility company may not have complete control over the water's distribution.

## CHINA

Within the boundaries of China lie some 1,500 rivers with basin areas greater than 1,000 square kilometers. Rivers with basin areas larger than 100 square kilometers number more than 5,000. The Chinese coastline extends for 18,000 kilometers, along which many rivers enter the sea.

There are 2,300 lakes in China with surface areas larger than one square kilometer. The total area is about 71,787 square kilometers; the total water storage volume is about 708,800 cubic meters, including 226,100 cubic meters of fresh water (equal to 31.9 percent of the total).

About 8,500 reservoirs, with a total storage capacity of 430 billion cubic meters, including 338 large reservoirs with a capacity of 306.8 billion cubic meters, have been built in China. However, due to wastewater from point and nonpoint sources into these reservoirs, putrification has become a serious problem, and some reservoirs have been converted from domestic water to wastewater reservoirs.

The Yangtze River is the largest river in China. It is more than 6,300 kilometers long, with a drainage area of 1.8 million square kilometers, yielding a mean annual runoff of about 1,000 billion cubic meters.

Both length and mean annual runoff rank the river third in the world. There are more than 350 million people living in the river basin area which accounts for one-third of China's total population.

The Yangtze has played an important role in the social and economic development of China. The Yangtze basin is one of the most important economic regions in China. The industrial and agricultural output value in the basin account for approximately 40 percent of the total in China.

The natural quality of the Yangtze water is good, both chemically and physically, making it suitable for various purposes. Along with the development of the economy in the basin area, however, huge amounts of domestic sewage and industrial wastewater are discharged into the Yangtze without proper treatment. According to a 1982 survey, the total wastewater discharged into the Yangtze was more than 12 billion tons per year, or 35 million tons per day, which accounted for 40 percent of the total in China. As a result, the water quality has deteriorated during the past decade and in some river reaches and lakes of the basin, water pollution has become quite serious.

In spite of these immense water resources, water shortage has become the main restricting factor for social and economic development. The water shortage is especially serious in large population areas in the north and in those regions where the economy is rapidly developing. Since the economic development was started around the lakes, large amounts of industrial and domestic wastewater have been flowing into the lakes and putrification has become more serious day by day.

Because the regional water resources are very limited, strengthened management and protection of water resources have become a main focus of the Chinese government to keep the ecological balance, to ensure balance between the supply and demand of available water resources, and to promote continued economic development.

## RUSSIA

The world's largest single freshwater source—Lake Baikal in the Russian border region near Mongolia—is being threatened by a new free market economy. One of the principal polluters is a paper and cellulose factory in Baikalsk, located at the southern end of the lake. Since 1966, the plant has poured an estimated two billion cubic yards of polluted water into the lake. Before the factory installed wastewater treatment equipment, the effluent contained such toxics as PCBs and sulfuric acid. In addition to industrial polluters, there reportedly is uncontrolled residential development along the giant lake's shoreline.

Russian officials are working with several international organizations, including the World Bank, to install more effective pollution controls along the Selinga River, which feeds Lake Baikal. They complain that businessmen, who claim land surrounding the lake, are thinking only about their profits. Some would like to see the paper plant shut down, but that seems unlikely to happen without suitable work for the factory's 3,500 employees.

## MEXICO

Only 76 percent of Mexican households have access to running water and only 61 percent have sewer connections. Despite progress made over the past three years by a "National Program of Potable Water and Sewage," less than 15 percent of municipal water in Mexico is treated.

Unrealistically low prices for water and overuse are a problem in Mexico City where there is currently no economic incentive for reducing water consumption. Despite the fact that 8 million Mexico City residents lack running water, the city uses 60,600 liters of water per second, while the wastewater load is 45,400 liters per second. The grim reality for Mexico City is that the closed groundwater basin which provides about 70 percent of the city's supply is polluted and there is not enough potable water. The volume of water Mexico City pumps from area aquifers exceeds recharge by nearly 13 percent. The great cathedral in Mexico City is sinking because of overdrawn groundwater.

Water sources for the metropolitan area are already exhausted and the city is dependent on imported water. The most expensive tap water is brought from the Cutzamala Valley to the north at 12,000 pesos per quart. This high price is mostly due to pumping costs, because Cutzamala is 3,600 feet above sea level and Mexico City is at an elevation of 7,400 feet.

The Coatzacoalcos River in southern Mexico, a 260-mile-long waterway that discharges into the Gulf of Mexico, has been described as one of the world's most contaminated waterways. Along its reaches, scores of communities and over 100 industries dump their sewage directly into the river. Pemex, the Mexican oil monopoly, adds a toxic brew of mercury, lead, sulfur, and phenol from its refineries. A tributary of the river contains 1,000 parts per million of mercury. The Teapa and Gopalapa rivers, tributaries of the Coatzacoalcos, are so polluted that dredging in the only method of cleanup. According to Pemex and other industries along the river, any sort of cleanup will be expensive and will likely not happen any time soon. A lone federal inspector in the area has time to check only the largest plants.

With the Mexican economy on the mend, improvement of water and drainage systems is a top government priority. A three-phase restructuring of Mexico City's water distribution, pricing and maintenance systems will eventually reduce water consumption by 20 to 40 percent and halt depletion of the aquifer underlying the city.

In order to finance water treatment improvements, the Mexican government is looking to private investors worldwide. The Mexican Investment Board is currently working with the World Bank and the Inter-American Development Bank to provide financial support to private investors. The board has targeted several water treatment projects to receive nonrefundable supports representing up to 35 percent of project costs. In its water supply strategy, the Mexico government is creating autonomous "Water Operating Units" that cross municipal and state jurisdictions to create districts that can more adequately handle existing water problems. To date, 142 operating units have been created.

## AUSTRALIA

A toxic algae bloom along some 600 miles of the Darling River in eastern Australia is being referred to as one of the greatest disasters of blue-green algae in the world. Some are calling it Australia's worst ecological disaster.

Inch-thick algae has plagued not only the Darling River but other waterways and lakes across Australia as well. The Australian government has established water purification facilities manned by the army along some stretches of the Darling, and wells are being drilled for new sources of drinking and irrigation water. Officials say that drought, sewage, and chemical fertilizers have all contributed to the unprecedented algae bloom.

Under the Constitution, the Australian Commonwealth has only limited direct powers in relation to water matters. The various states of Australia also deal with water related construction and other operational matters. The State responsibilities include water resources assessment, flood mitigation and river management. Thus, the approach to water monitoring varies from state to state as does the legislation pertaining to water quality.

These examples (which are by no means inclusive) illustrate that, in many countries of the world, there is still a vast potential for developing their available water resources. In many countries, however, more than one agency is responsible for water resources management. This often leads to the problem of overlapping and lack of coordination. Environmentally sound development and management of these water resources is essential in order to achieve sustained socioeconomic growth in both developing countries and industrial countries.

Earth's exploding population growth, along with its resultant industrial and domestic water use, requires water resource managers to have in place long-range water supply plans. Water resource managers must continue to plan for additional water supplies, as well as take measures to stretch and more efficiently use current supplies.

Finally, there is a need for increased international assistance in terms of financial resources, equipment, technical know-how, training of technical and management personnel etc., in order to establish new and/or strengthen existing water resources management systems.

## REFERENCES

*Water Quality Monitoring in the Asian and Pacific Region, Water Resources Series*. No. 67, United Nations, New York, 1990

Kuhn, Wolfgang. A Unifying Perspective on Water Supply. *AWWA Mainstream*, November 1992, Volume 36, No. 11, Pg. 2.

Bourbigot, Marie-Marguerite. Research Plays Practical Part in Water Supply. *AWWA Mainstream*, May 1993, Volume 37, No. 5, Pg. 2.

Lake Baikal Threatened by Free Market Pollution. *U.S. Water News*, March 1993, Volume 9, No. 9, Pg. 2.

Reid, Joe. WFP Volunteers Visit Poland. *AWWA Mainstream*, February 1993, Volume 37, No. 2, Pg. 7.

*The Water Encyclopedia*. Van der Leeden, Frits. Lewis Publishers, Inc., Chelsea, Michigan, 1990.

River in Mexico Called One of World's Most Polluted. *U.S. Water News*, December 1992, Vol. 9, No. 6.

Mexico City Begins Overhaul of Water Supply System. *U.S. Water News*, November 1992, Vol. 9, No. 5.

Mexico Seeks Investors for Water Treatment Improvements. *U.S. Water News*, May 1993, Vol. 9, No. 11, pg. 2.

# CHAPTER ONE
# Water-Demand Characteristics

The first task facing the water planner is to choose a supply philosophy that reflects the level of risk the consumer is willing to accept. No water system—no matter how conservative its design—can meet every conceivable demand condition. Therefore, every water manager must deal with risk whenever ultimate demands are projected.

Traditionally, risk has been expressed as a measurement of reliability, with a 98 percent success rate often cited as an acceptable level of risk (Miller and Ludlum 1985). In other words, 98 percent of the time, or in 98 out of 100 years, the supply source in question (surface, groundwater, or reclaimed water) must be able to deliver the capacity that has been ascribed to it by the water planner. Reliability standards vary from agency to agency, depending on the costs involved and other criteria. Sometimes conservative water-use factors and/or design criteria are used in place of quantifications of the level of success achieved to ensure a high degree of reliability.

The planning and design criteria used for storage, treatment, and distribution facilities will reflect the supply philosophy chosen by the water utility. Such criteria are based on observed variations in short- and long-term water demands. In choosing a supply philosophy, the water planner must determine the impact of demand variations on the forecasting process. The issue of regionalization (joint use of facilities) must also be addressed.

## WATER-DEMAND CRITERIA

### SPATIAL AND TEMPORAL VARIATIONS IN WATER DEMAND

Urban water demand is usually expressed as a function of population and per-capita consumption (in gallons per capita per day, or gpcd). Forecasts of average daily per-capita municipal and industrial demand provide a useful measure of the mean quantity of treated water necessary to support domestic water use, as well as of an individual's proportionate share of the water consumed by supportive industrial, commercial, and municipal uses (Orange County Municipal Water District 1985). For new areas, the per-capita use factor is usually based on an average of observed water use rates in existing communities with comparable development characteristics located in similar climatic regions.

Per-capita water consumption is highly variable from area to area and may even vary on a local basis. Per-capita consumption is inversely related to rainfall because outside irrigation use declines during wet periods. Hot dry areas have especially high landscaping water demands, and these increase the per-capita use; whereas the higher humidity and moderate temperatures of coastal areas decrease outside water demand and reduce per-capita consumption. Besides being related to rainfall, differences in consumption rates are related to the following causal factors:

- Variations in lawn-irrigation demands associated with differences in residential density
- Differences in greenbelt irrigation requirements and in the availability of untreated or reclaimed water for these needs
- Differences in the degree to which structural and nonstructural water-conservation measures have been implemented in the area
- Variations in the persons/household ratio
- Variations in the concentration of water-intensive industrial and commercial land uses
- Effectiveness of public-education programs to increase consumer awareness
- Variations in income levels and other economic criteria
- Intensity of construction activity, such as grading and site work

In the short term, water demand fluctuates seasonally (monthly), daily, and hourly. These fluctuations in demand are normally estimated in terms of peaking factors, which are expressed as a percentage of average annual demand. In general, seasonal peaking is used to size regional supply facilities, while daily peaking and hourly peaking are used to size local distribution facilities.

## Seasonal Peaking

Maximum-month demands are used to determine the capacity requirements of major water-supply facilities, such as regional transmission pipelines and regulatory storage reservoirs. Regulatory storage reservoirs are usually sized to provide a supplemental source of water during the hot summer months, when demands may exceed the capacity of base-load supply facilities (transmission pipelines, water-treatment plants, groundwater pumps, and so on). In the absence of seasonal storage, base-load facilities must be able to meet the full maximum-month demand.

Supply capacity is traditionally sized to accommodate a maximum-month demand that varies between 150 and 200 percent of the average daily demand. In the Northern Hemisphere, seasonal peaking is highest in the months of June through September. Figure 1-1 shows a typical hydrograph of seasonal water demands for an urban water system.

## Daily Peaking

Maximum-day demands are used in conjunction with peak-hour demands and fire-flow requirements to size local distribution facilities (such as storage tanks, booster-pump stations, and distribution pipelines) in the water system. The most critical demands involve the sizing of system piping and service pumps in relation to maximum-day demand, maximum storage-replenishment rate, and maximum-day demand plus fire-flow demand; a second critical area involves the sizing of system storage, booster pumps, and service pumps in relation to maximum-day demand plus fire-flow demand; finally, sizing for peak-hour demand is essential (Topping and Puccia 1984).

Daily peaking varies considerably from one system to another, depending on the water-using characteristics of the service area. Daily peaking can also be affected by how each system is operated. If water-service elevations in operational storage reservoirs are monitored closely and if regional supply deliveries are changed frequently, daily peaking can be controlled. Where close attention to detail is neglected, however—as in a case where copious operational storage is available to accommodate supply deliveries in excess of actual demands—daily peaking can be excessive. In this situation, water system operators will accept higher-than-needed deliveries until all storage space is used up; this results in an artificially high peaking rate. In most cases, maximum-day demands are around 200 percent of average daily demand.

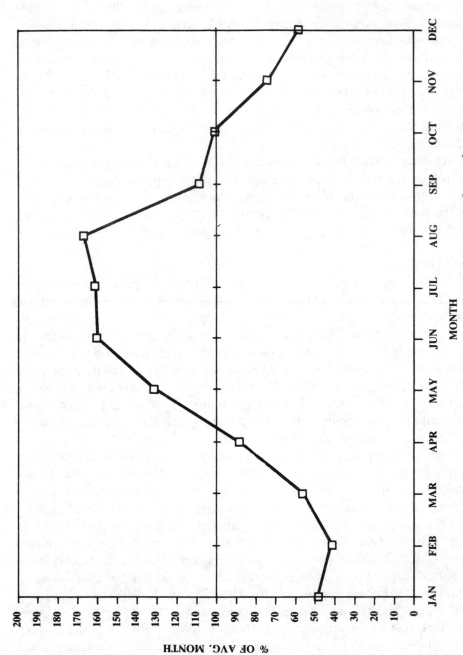

Figure 1-1. Typical seasonal hydrograph: urban water demand.

## Diurnal Peaking

Peak-hour demands are used in designing water-distribution systems. Operational storage facilities and the hydraulic capacity of a distribution system must be capable of supplying the peak-hour demand. Peak-hour demands can be measured in the field by isolating one part of the service area and measuring all inflows and outflows during a maximum-day period. The resulting diurnal hydrograph uniquely reflects the water-using characteristics of the service area. The fluctuations in demand recorded during an assumed maximum-day period may not represent the greatest possible degree of peaking, however; consequently, peak-hour projections often use a conservative (high) peak hour/annual average water-use ratio. Distribution systems are traditionally designed to carry peak-hour flows that amount to between 200 and 300 percent of the average day demand, with higher values usually associated with smaller systems (Robinson and Blair 1984).

A review of diurnal water use for large metropolitan areas having a normal complement of industrial and commercial facilities indicates that municipal water demand typically experiences two peaks during a maximum-day period. The first peak occurs at around 7 to 9 A.M., and the second (and usually larger) peak occurs at around 7 to 9 P.M. The magnitude of the peak varies considerably from one community to another and according to the day of the year.

Among the few studies of diurnal peaking is a study by Wolff, Linaweaver, and Geyer (1975) of the Baltimore metropolitan area. Figure 1-2 shows a typical diurnal hydrograph for residential areas of Baltimore during winter and summer maximum-day periods. The marked difference between winter and summer peaking can be attributed to residential lawn sprinkling. In the Wolff study, diurnal peaking was recorded for eighteen different types of commercial and institutional establishments by continuous readings and visual observations of water usage. The observed data on peaking are summarized in Table 1-1. With only two exceptions, the peak usage occurred between 8 A.M. and 6 P.M.

Wolff, Linaweaver, and Geyer used the observed peaking patterns to develop a synthetic hydrograph for a "typical" residential community of 100,000 people (28,600 family units) that has a normal complement of industrial and commercial facilities serving it (fig. 1-3). For a typical winter day (with no lawn sprinkling), the commercial demand during

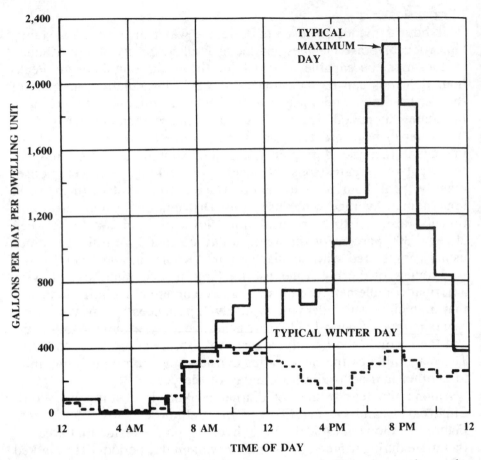

Figure 1-2. Hourly residential water use on a typical maximum day and a typical winter day. (Reprinted from Wolff, Linaweaver, and Geyer 1975, p. 56, by permission of ASCE)

the peak hour (10 A.M.) constituted 20 percent of the peak demand. On a hot summer day (with sprinkling), the commercial demand during the peak hour (8 P.M.) constituted only 4.4 percent of the peak demand. The researchers concluded that commercial demand does not materially increase peak-hour demand during periods of maximum water use, as a result of its favored location in the demand hydrograph.

**Table 1-1. Summary of Commercial and Institutional Diurnal Peaking.**

| Type of Establishment or Institution | Hour of Peak Occurrence | Type of Establishment or Institution | Hour of Peak Occurrence |
|---|---|---|---|
| Primary and secondary schools | | Hotels | 6 P.M.–7 P.M. |
| public elementary | 1 P.M.–2 P.M. | Motels | 10 A.M.–11 A.M. |
| public senior high | 4 P.M.–5 P.M. | Office buildings | |
| private elementary | 10 A.M.–11 A.M. | general offices | 12 noon–1 P.M. |
| | | medical offices | 11 A.M. and 4 P.M. |
| private senior high | 4 P.M.–5 P.M. | Department stores | 12 noon–1 P.M. |
| combined (grades 1–12) | 5 P.M.–6 P.M. | Shopping centers | 2 P.M.–3 P.M. |
| | | Car washes | 12 noon–2 P.M. |
| Colleges | | | |
| students in residence | 1 P.M.–2 P.M. | Service stations | 12 noon–1 P.M. |
| nonresident students | 8 A.M.–9 A.M. | Laundries | 2 P.M.–3 P.M. |
| | | Restaurants | 2 P.M.–3 P.M. |
| Hospitals | 11 A.M.–12 noon | | |
| | | Churches | 8 P.M.–9 P.M. |
| Nursing homes and institutions | 4 P.M.–5 P.M. | | |
| | | Barber shops | 2 P.M.–3 P.M. |
| Apartments | | Beauty salons | 2 P.M.–3 P.M. |
| high-rise | 11 A.M.–12 noon | | |
| garden-type | 8 A.M. and 6 P.M. | | |

SOURCE: Adapted from Wolff, Linaweaver, and Geyer (1975), p. 49, by permission of ASCE.

## WATER-FORECAST APPLICATIONS

Having observed monthly, daily, and hourly fluctuations in demand, the water planner can adjust the demand projections to fit various applications. Often the intended application of the forecast is a primary consideration in selecting a forecast method (Boland et al. 1983). For example, if the results will be used to evaluate the long-term adequacy of alternative supply sources, a forecast of average annual aggregate water use is sufficient. If the intended application involves sizing a

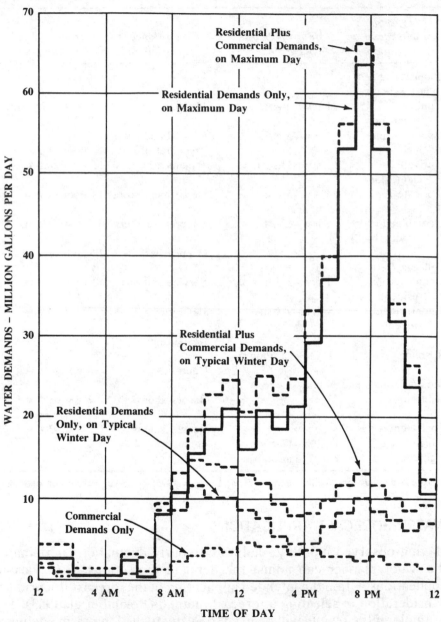

Figure 1-3. Synthetic Hydrograph for a community with a population of 100,000, in 28,000 owner-occupied dwellings. (Reprinted from Wolff, Linaweaver, and Geyer 1975, p. 57, by permission of ASCE)

facility so that it can handle seasonal or short-term peak demands, a disaggregate forecast by type of land use may be required. For example, the design of a surface-water reservoir may (depending on project size and purpose) require forecasts of seasonal water use and maximum-month use, as well as of average annual water use.

Designing covered operational storage facilities may require forecasts of maximum-week or consecutive maximum-day water use. If the forecast is to be used in designing treatment and conveyance works, a reliable estimate of maximum-day water use may be needed. In some systems, pump station design may ultimately require forecasts of maximum-day or even maximum-hour water use at ultimate development. Such forecasts, depending on the degree of accuracy required, may call for methods involving multiple-use coefficients, probability analyses, and other complex techniques. These forecasting methods are discussed in chapter 2.

## REGIONALIZATION

After identifying relevant short- and long-term water needs, by delineating appropriate consumption rates and peaking factors, the water planner should address the issue of regionalization of supply facilities. The American Water Works Association (1986) defines *regionalization of water systems* as "creation of an appropriate management or contractual administrative organization or a coordinated physical system plan of two or more community water systems in a geographical area for the purpose of utilizing common resources and facilities to their optimum advantage."

The concept of regionalization has evolved from the historical development of water utilities, which have themselves developed where populations become concentrated. Except for systems that serve the very largest cities, water systems are small in comparison to other utilities. There are more than 60,000 water systems in the United States (Johnstone 1985); and according to estimates by the U.S. Environmental Protection Agency, slightly more than 43,000 of these systems serve only between 25 and 500 persons each (Shelstad and Hanson 1986). Most water systems were intended to be self-sufficient. The cost of laying pipelines to traverse sparsely populated areas has militated against the development of interconnected, regionalized systems. Small water systems include public systems serving small towns and villages; privately owned systems serving institutions, mobile-home parks, and subdivisions; and "mom-and-pop"-type private operations serving small communities.

## Problems Faced by Small Water Systems

According to Shelstad and Hanson (1986), small systems often face problems of inadequate capital, inadequate water reserves, and inadequate rates. These problems are due primarily to a lack of economy of scale—since capital, operation, and maintenance costs cannot be spread over enough customers to keep individual costs low—and to the absence of a dependable, growing customer base from which to finance capital improvements.

Financing capital improvements has always been a problem for water utilities because the water industry is capital-intensive. Unlike other utilities, water utilities have little opportunity for cost reduction through technological change, since most water systems facilities have long lives. A substantial increase in investment occurs when plants built many years ago at relatively low construction costs are replaced at today's higher costs and higher interest rates. As described by Johnstone (1985), "That distribution main installed many years ago at three dollars a foot and four percent money will be replaced today at twenty dollars a foot using ten percent money." Another problem is that new facilities built to increase system capacity usually cost more per unit of output because new sources of supply are farther away, more costly to develop, and (often) of poorer quality.

As a result of their special problems, 15 percent of the very small, investor-owned water utilities in the United States are unable to finance routine expenditures, according to an EPA survey (Shelstad and Hanson 1986). The culprit is often inadequate rates. In some cases, operators of small systems fail to keep detailed records on costs that could have supported a petition to a regulatory agency for higher rates. Typically, such operators approach the regulatory agency infrequently—although when they do, it is to request a substantial increase in rates. The net effect of this practice is to keep rates artificially low; the same result may occur for political reasons, or water-system revenues may be used to subsidize other municipal programs instead of to pay utility costs. Privately owned systems are subject to the rate-making decisions of public utility commissions, which often feel considerable public pressure to keep rates low. Larger systems, in contrast, may avoid much of the political fallout by making annual requests for small, incremental increases.

## Types of Regionalization

Regionalization may take place under two distinct circumstances (Coelen 1981). First, two or more existing systems may share production facilities without an extension of service. Alternatively, service may be extended from one or more existing systems into areas previously lacking service. These two situations differ in respect to the way that project benefits are calculated and the way that those benefits are distributed over the regional population.

In the situation involving no extended service, all systems participating in the regionalization benefit from the increased economy of scale. The resulting savings, either to the consumer or the municipality, can be spent on other valued consumption. Other, more complicated situations may also be comprehended. For example, a water-rich system may operate without regional interaction during normal times; then, in times of drought (or other emergency), it may transfer water to its water-short regionalized partner. This situation shows how regionalization may serve analogously to an insurance policy, with the impetus for regionalization being the sharing of risks.

Because not all water systems abut other systems, many cannot share facilities. In this situation, extended service provides the opportunity to expand the service area by acquiring adjacent systems or by encouraging other systems or industries to discontinue use of their supply and instead become customers (Johnstone 1985). Where this has occurred, the benefits of service extension primarily consist of the increased value of water supply consumption. This is likely to be reflected by increasing property values in areas with new customers.

The key element involved in assessing the value of regionalization is the ability of the regionalized partners to share (and thereby lower) risks. Risk reduction is possible only under certain circumstances. For example, risks can be lowered only if the regionalized partners draw their supplies from hydrologically different areas. In view of this prerequisite, the benefits of risk-sharing may be offset by high regionalization costs due to the extended distance between potential partners of different hydrologic areas. Again, if at least one system in a regionalization network has substantial surplus supplies available at a reasonable price, water-poor partners may enjoy a significant risk reduction; and yet regionalization prospects may still be poor because the water-rich area has nothing to gain from regionalization unless it can sell its water to the water-short areas at such high prices that the neighboring systems will balk at regionalization.

The greatest potential for risk reduction through regionalization is in an area where new supplies must be generated to serve as a backup for drought periods. Given the expensiveness of the investment, small suppliers who are prevented by regulation from accumulating capital may not be able to finance individual projects. Acting together as a regional system, they may form a sufficiently large bloc to obtain necessary capital funding through loans or bonds. In addition, even if the suppliers could obtain capital individually, the regional risk-serving water supplies could probably be developed at lower average cost than could the individual supplies.

## Requirements for Effective Regionalization

Johnstone (1985) cites the following requirements for effective regionalization:

1. *Strong institutional arrangements to surmount local and regional jurisdictional barriers*. A simple and easily understood contractual arrangement must be established.

2. *Assignment for costs for joint use of existing facilities on a fair and equitable basis*. In cases where regionalization is warranted because benefits outweigh costs, projects often fail to be adopted because the benefits are distributed to only a few individuals while the costs are spread among many. Everyone should pay less for water service under regionalization than they would otherwise. In some cases, the next increment of supply may be postponable by maximizing the use of existing facilities.

3. *Economic responsibility must be properly assigned to customer groups*. The failure to develop adequate sources of water over a large region can cause water-supply restrictions in every system—including ones that made use of proper planning and source development. Poor planning, leading to inadequate supply and the need for outside help, must not be rewarded. The customers of systems that are under stress might be enjoying artificially low water rates because their system does not have sufficient supply currently

in place. A utility that has planned ahead must not be penalized by being forced to incur higher costs in supplying others in need and then having to pass these costs on to their own customers.

## Future Trends in Regionalization

Shelstad and Hanson (1986) predict that many small water systems will be forced to merge with larger systems as a result of the rising cost of compliance with increasingly stringent regulatory standards. Some states have already taken action either to force consolidation or to encourage voluntary cooperation among water utilities. The following three cases are illustrative:

**1.** New Jersey has enacted a law designed to enforce compliance by utilities with statutory and regulatory requirements relating to water quality, pressure, and volume. A takeover, by either a public or private entity, may be ordered for any small company not in compliance with these standards.

**2.** Connecticut is envisaging dividing the state into water-supply management areas—clusters of towns with related water-supply concerns. Committees representing the local water utilities would work together to develop coordinated plans, with incentives encouraging the takeover or satellite-management of small water companies that are inefficient and ineffective. In cases where voluntary cooperation would not be successful, Connecticut's Water Resources Task Force has recommended legislation to put small companies into receivership when their problems are attributable to neglect or abandonment.

**3.** The state of Washington's Public Water System Coordination Act of 1977 has established a step-by-step process for planning a satellite support system. In a satellite system, ownership may be transferred to a larger utility or umbrella organization, but small utilities can choose from among many other options and can contract with a larger utility or umbrella organization for such services as repairs, system operation and maintenance, monitoring and reporting, administration and billing, and wholesale water supply.

Johnstone (1985) foresees a less-drastic situation in which interconnections of adjacent water systems are statutorily required. Through these interconnections, water will be shared during emergencies, including drought. Cost responsibility for constructing interconnections and paying for the water transferred will have to be determined by regulatory agencies using predetermined guidelines. Further, in areas where existing sources are fully utilized, it may be necessary to shift a water system from one source to another, simply to allow another utility to use the first source. Clearly, this might force a group of nonaffiliated customers to substitute a higher-cost supply for a lower-cost supply.

Forced consolidation is an expensive legal process that is appropriate only as a last resort. Attempts to force consolidation have met with considerable opposition from water customers, who fear that their interests will be neglected by larger utilities, and from private utilities concerned with their property rights. Except in hopeless cases, consolidation should not be imposed from the top down; instead, it should be achieved through a process of voluntary cooperation.

## WATER-SUPPLY PLANNING PHILOSOPHIES

After examining demand fluctuations, the water planner chooses a forecast method and a technique commensurate with the level of accuracy dictated by the use of the forecast results. Where feasible, regionalization options designed to minimize the risk of water shortages are explored. At this point, the water planner can begin formulating a planning philosophy for the design of facilities to accomodate ultimate demand conditions. Formulating such a philosophy presents an opportunity to manage the risks inherent in the forecasting of water demands.

Water-supply facility planning involves evaluating the capacity of alternative water-supply sources to meet projected demands. While most of the emphasis in evaluating alternatives is on cost (including basic financial feasibility), other important criteria must also be evaluated—such as environmental impact, regional benefits, water quality, and practical construction feasibility. It is essential to use realistic planning criteria, since municipal water demands tend to fluctuate greatly, especially over short periods of time. Millions of dollars in facility costs may be riding on the definition of a single planning criterion.

Planning criteria can be generalized into two water-supply planning philosophies. In accordance with the first of these, facilities are designed and constructed to deliver a specified continuous level of supply under all conditions, regardless of cost. Under this philosophy, a conservative set of assumptions is used to define future conditions affecting water demand. Water-capacity requirements are then based on the resulting ultimate demand projections. Joint-venture options are rejected in favor of complete control by the utility over its sources of supply and its system operation. These factors minimize the risk of inadequate water service, but they also tend to yield projects with excess capacity and hence unnecessarily high costs. Given low interest rates and construction costs, however, the impact on water rates may be acceptable; and in areas of rapid growth, the capacity is likely to be needed eventually.

Under the second philosophy, facilities are designed and built to deliver something less than a specified continuous level of supply. Water-capacity requirements are instead based on a system that will periodically require varying levels of consumer cooperation to maintain a supply buffer during extremely dry years. The philosophy also relies on a willingness to invest money in facility-sharing options (regionalization), structural efficiency, leak detection, and reclamation; and it may entail consumer costs in the form of retrofitting, brown lawns, and inconvenience, in order to avoid developing substantial new supplies that may or may not be needed. Such a risk-management method involves a balancing by the forecaster of appropriate risk level on the one hand and an acceptable "sacrifice level" on the other. Implicit in this method is the assumption that people's behaviorial and structural conservation life-styles can be estimated—an issue that is discussed in detail in chapter 2.

The first planning philosophy is based on the traditional factors used by water managers to evaluate new projects: economics and operational feasibility. The second planning philosophy also considers these factors, but it also includes the perceived attitudes and expectations of its customers.

An example of the impact of a changing water-supply planning philosophy is the experience of the Marin Municipal Water District (MMWD) following the 1976–77 drought. Located in Marin County, California, the 28-square-mile, district-owned watershed obtains approximately 95 percent of its supply from rainfall. As a result of a voluntary conservation program adopted during the drought, total annual water consumption decreased from 32,300 to 26,300 acre-feet, a reduction of about 22 percent (Teknekron Research 1979). Following the drought, Marin planners assumed that a conservation carry-over would occur, in addition to conservation from measures being enforced in new developments (by requiring low-flow devices, lowered water pressure, and efficient irrigation systems) and conservation from an intensive leak-detection program. The resulting projection of future water demands is considerably lower than a projection based on a continuation of historical trends (fig. 1-4). The effectiveness of this conservation carry-over is reflected in the area's postdrought consumption levels. At the end of fiscal year 1979–80, the level of per-capita water consumption in Marin County was still 22.5 percent below the predrought level. Actual consumption for 1979 was one-half the predrought projection for that year (Woo 1982).

## WATER-DEMAND COMPONENTS

In large metropolitan areas encompassing a variety of service area characteristics, water demand should be forecast on the basis of a more sophisticated method of evaluation than simply population times a composite historical consumption value. In areas where significant growth is occurring, fine-tuning of per-capita use estimates will be overshadowed by the uncertainty of the population projections. For areas approaching saturation population, the use of disaggregate forecasts may reveal trends suggesting ways to use water resources more efficiently. Flexibility is desirable in establishing planning goals to address variations from projected demands. In particular, facility planning should take into account the type of land use, as well as other appropriate factors, in a number of smaller areas within the total service area.

As per-capita water-use forecasts grow less common, the need to examine the characteristics of the major components of urban water demand will increase. Urban water demand can be divided into two major categories: residential demands, and nonresidential demands. Nonresidential demands, in turn, can be subdivided into industrial demands, commercial demands, unaccounted-for uses, and other components.

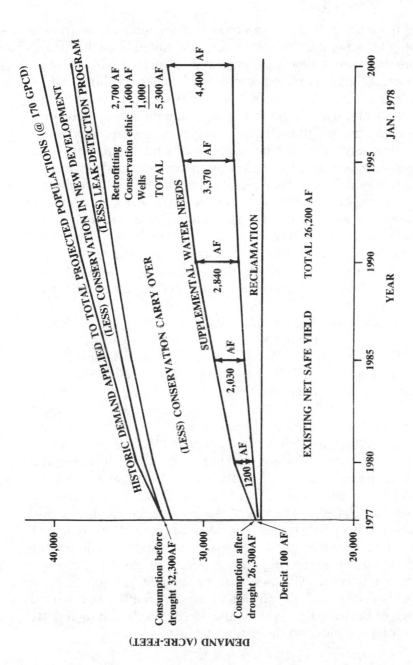

Figure 1-4. Marin's projected supplemental water needs, 1977–2000. (Reprinted from Marin Municipal Water District, *Environmental Assessment—1977 New Water Supply Alternatives*, January 1978)

## RESIDENTIAL DEMANDS

Residential demands consist of an indoor component and an outdoor component. The percentage of total water use found in each category depends on the water-using characteristics of each housing-density level. In general, large families use more water per dwelling unit but less water per capita; the size of the family does not significantly affect outdoor water use. Single-family dwelling units tend to be more water-intensive than multifamily units because they are responsible for a proportionately larger amount of outdoor landscaping per capita. Indoor water use is usually assumed to be constant per capita for both single-family and multifamily units, but the same assumption is not made with regard to outdoor water use. Affluent customers, for example, may use more water for landscaping large outdoor areas. In addition, water consumption is usually higher in sewer-equipped communities than in areas where septic tanks are used.

In humid areas and in apartment buildings, most water is used inside the home. In average residential areas located in an arid climate, about half the water is used inside the home, and the rest is used outside for landscaping. Figure 1-5 shows the percentage breakdown of interior water use compiled by the U.S. Department of Housing and Urban Development (HUD) as part of a nationwide survey conducted in the early 1980s (HUD 1984). Breakdowns of interior water use are shown for both nonconserving and conserving homes (homes equipped with water-saving devices). As shown in figure 1-5, most of the interior water use occurs in the bathroom.

Short-term fluctuations in indoor water demands can be measured from sewage-flow records. Indoor demands exhibit very little seasonal peaking, since they are not significantly affected by climatic changes. Typically (based on an examination of sewage-flow records), seasonal variations do not exceed 10 percent of the average annual flow. Short-term fluctuations in a predominantly residential community are usually due to variations in water use for landscape irrigation. The amount of water spent on landscape irrigation depends on climatic influences and irrigation efficiency. Climatic factors include precipitation, air temperature, humidity, and air movement. Irrigation efficiency depends on the design of the irrigation system and on soil profile characteristics related to root-zone moisture-holding capacity.

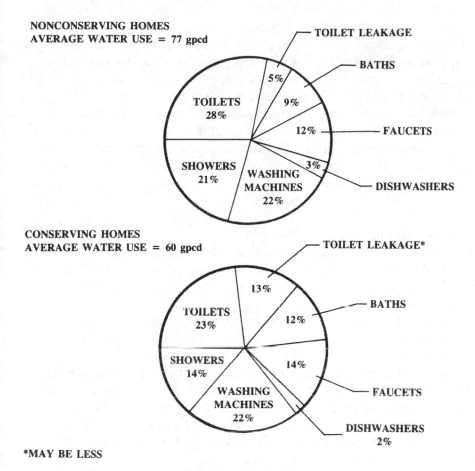

NONCONSERVING HOMES
AVERAGE WATER USE = 77 gpcd

TOILET LEAKAGE

BATHS

TOILETS
28%

5%

9%

12% — FAUCETS

SHOWERS
21%

WASHING
MACHINES
22%

3%

DISHWASHERS

CONSERVING HOMES
AVERAGE WATER USE = 60 gpcd

TOILET LEAKAGE*

BATHS

TOILETS
23%

13%

12%

SHOWERS
14%

14%

WASHING
MACHINES
22%

FAUCETS

DISHWASHERS
2%

*MAY BE LESS

Figure 1-5. Comparison of per-capita interior water use for nonconserving and conserving homes. (Reprinted from HUD 1984, p. 1-3)

Urban development in recent years has been marked by an increase in townhouse and condominium construction. Outdoor irrigation in these multifamily developments is usually controlled by homeowner associations, which are maintained jointly by all residents. Areas under association control (collectively termed *common-lot areas*) include arterial slopes, greenbelts, median strips, and some building exteriors. The total acreage under the central control of these associations could represent a substantial portion of the total residential demand, especially in planned communities, where associations may also be responsible for maintaining community and regional parks.

By far the greatest factor affecting short-term peaking of landscape irrigation in common-lot areas is the psychological reactions of those responsible for operating the irrigation system. Operators of landscape irrigation systems tend to overirrigate; stuck valves and/or inflexible irrigation scheduling lead to irrigation even during rainy periods, with a resulting rise in consumer complaints. The magnitude of peaking observed in any system depends on the climate, the extent and character of the landscaping, and the operational mode of the irrigation system. Typically, water used for landscape irrigation has a seasonal peak (maximum-month) water use of around 200 percent of the average daily demand.

## Forecast Applications

In order to forecast future residential water demands, the water planner must identify and isolate the major areas of water use. The most important distinction to make here is between indoor and outdoor water use.

Per-capita consumption is inversely related to rainfall because outside irrigation is reduced during wet periods. During prolonged periods of rainfall, when the soil is saturated, outdoor landscape irrigation is virtually nonexistent. Consequently, during these periods, the indoor component of domestic water consumption can be taken as the entire amount recorded in monthly water billings. Typically, unit residential consumption rates during months of heavy rainfall vary from about 50 to about 100 gpcd. These consumption factors can also be checked with observed wastewater-flow data.

In some cases, a correlation between rainfall and housing density may be difficult to obtain because of such complicating factors as bimonthly billing cycles, end-of-year accounting adjustments in water billings, inaccurate water meters, and unaccounted-for water uses. In instances of this kind, adjustments would have to be made to arrive at a time-phased residential demand that reflected actual water consumption. The impact of inaccurate meters and of unaccounted-for water use are discussed later.

## INDUSTRIAL DEMANDS

In the United States, the largest withdrawal uses of water resources are for thermoelectric power generation, for irrigation, for manufacturing, and for municipal water supply. Among energy-generating util-

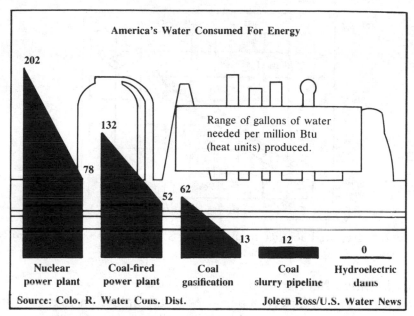

Figure 1-6. Water consumption in energy production, by industry. (Reprinted with permission of *U.S. Water News*)

ities, nuclear power plants consume the most water per unit of heat produced (fig. 1-6). Irrigated acreage totaled 44.6 million acres nationwide in 1983—an increase of about 9.5 million acres over the 1973 figure (fig. 1-7). Total on-farm pump irrigation grew by 27 percent during the same ten-year period (*U.S. Water News* 1986). Manufacturing industries in the United States now withdraw about twice as much water as do domestic water systems.

The source of water for small industrial facilities is the local water utility. Larger users, especially those located outside urban areas— canneries, refineries, pulp and lumber mills, and so on—tend to develop their own supplies from wells or from local stream diversions, resulting in a substantial amount of water use that is not reported in water-utility records (California Department of Water Resources 1982). A by-product of this situation is that much of the growth in water use projected for the industrial sector is limited to industries located in metropolitan and surrounding areas, where public water supplies are prevalent. For practical purposes, municipal water systems will be the source of all water used by city-located industries (Kollar and Brewer 1980*a*).

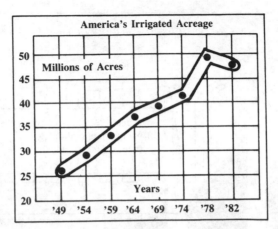

Figure 1-7. Irrigated acreage in the United States, 1949–82. (Reprinted with permission of *U.S. Water News*)

## Types of Industrial Demand

Water is used for a variety of purposes by the industries that constitute the manufacturing sector of the United States. In general, water use can be broken down into the following categories:*

| | |
|---|---|
| Cooling and condensing | 64% |
| Process | 33 |
| Air conditioning | 3 |
| TOTAL | 100% |

**Cooling Water (noncontact)**   In this application, heat-exchanger surfaces stand between the water and the material being cooled. The most common noncontact cooling processes are equipment cooling, process temperature control, and steam-electric power condensing.

**Process Water**   Included in this category are a wide variety of applications in which water comes into contact with process materials or waste products or is incorporated into the product. Common process applications are water inclusion in food and beverages, slurrying, paper-forming, spray-cooling, and barometric condensing. Boiler-feed

* This breakdown of industrial water use is adapted from Kollar and Brewer (1980b). The percentages have been modified to fit definitions of the components of industrial water use contained in the California Department of Water Resources (1982).

water is also included in this category; it is used to generate steam for process purposes or for steam-electric power generation.

**Air Conditioning**   In this application water is used as a heat-exchange medium in an apparatus for controlling the humidity and temperature of air.

### Standard Industrial Classification System

The standard practice of business statisticians in the United States is to deal with groups of industries by aggregating industry subsectors on the basis of similarities in their processors, raw materials, and final products. Two examples of aggregated industry subsectors are all food processing plants and all petroleum refineries. Each grouping is well defined under the Standard Industrial Classification (SIC) system.

The SIC system is an industrial classification of the entire economy. As such, it divides all activities into broad industrial divisions and then subdivides each division into two-digit major groups. Table 1-2 summarizes major manufacturing classifications (two-digit groups) contained in the SIC system. The SIC system also classifies the major (two-digit) groups into three-digit industry subgroups, and further separates these into four-digit detailed industries. The numbering system thus permits the use of classifications at various levels of detail—four-digit, three-digit, or two-digit—according to the specific uses desired.

For example, major group 37 (transportation equipment) includes establishments that manufacture equipment for transportation of passengers and cargo by land, air, and water. The following three-digit industry subgroups are included in this major group: motor vehicles and motor-vehicle equipment (371); aircraft and parts (372); ship and boat building and repairing (373); railroad equipment (374); motorcycles, bicycles, and parts (375); guided missiles and space vehicles and parts (376); and miscellaneous transportation equipment (379). The following four-digit detailed industries are included in subgroup 371 (motor vehicles and motor-vehicle equipment): motor vehicles and passenger-car bodies (3711); truck and bus bodies (3713); motor-vehicle parts and accessories (3714); and truck trailers (3715).

The SIC system can be manipulated to serve the water planner's own special needs. For each study area, certain major industrial groups are readily identifiable as high-volume water users; each of these must

**Table 1-2. Manufacturing Classifications Used in the SIC System.**

| Major Group Number | Classification | Major Group Number | Classification |
|---|---|---|---|
| 20 | Food and kindred products | 32 | Stone, clay, glass, and concrete products |
| 21 | Tobacco manufacturers | 33 | Primary metal industries |
| 22 | Textile mill products | 34 | Fabricated metal products, ordnance, machinery, and transportation equipment |
| 23 | Apparel and other finished products made from fabrics and similar materials | | |
| 24 | Lumber and wood products, except furniture | 35 | Machinery, except electrical |
| 25 | Furniture and fixtures | 36 | Electrical machinery, equipment, and supplies |
| 26 | Paper and allied products | | |
| 27 | Printing, publishing, and allied industries | 37 | Transportation equipment |
| 28 | Chemicals and allied products | 38 | Professional, scientific, and controlling instruments; photographic and optical goods; watches and clocks |
| 29 | Petroleum refining and related industries | | |
| 30 | Rubber and miscellaneous plastic products | | |
| 31 | Leather and leather products | 39 | Miscellaneous manufacturing industries |

SOURCE: Office of Statistical Standards (1972).

be examined in detail. Thus, industries may be grouped according to their water requirements and their potential impact on local water sources and suppliers. Because variable-rate water users appear even in the same four-digit listing, subdividing some of the detailed industries at this level is extremely useful. For example, the use of different processes to manufacture the same product may cause a significant variation in the water requirements (per unit of product) of two different manufacturers. Therefore, the SIC system should be refined by adding one and possibly two digits to the four-digit level of industry identification, to enable the water planner to distinguish between significantly different water-using entities engaged in producing essentially the same product. As time goes on and as more information is obtained about identifiable differences in similar industries, these differences can be documented and the SIC system extended to account for them.

## Industrial Water Recycling

The cost of water is relatively insignificant in relation to other manufacturing expenses, so cost is not an appreciable deterrent to its use. A report of the National Water Commission (1973) states that the cost of water supply for industry is usually less than 2 percent of production costs. The cost of water has risen somewhat in recent years, but not as sharply as have the costs of discharging large volumes of untreated wastes. New federal and state controls have substantially increased waste-discharge costs, and for this reason the incentive to conserve water has grown. Dischargers must pay the cost of treatment in public treatment facilities and must provide their own wastewater treatment when discharging into navigable receiving waters or into areas where groundwater might become contaminated. Thus, water conservation becomes essential where substantial quantities of water are required for processing. To compete in today's market, many industries are compelled to reduce their intake of fresh water, reduce their waste loadings, recycle water, and use treated wastewater.

By definition, industrial water recycling is the direct reuse of water, without treatment or with limited treatment, at the same general location or for the same purpose. Reclaimed water is water that has been treated in a reclamation plant and is usually transported to another location for reuse. A manufacturing plant may reuse its own wastewater, with or without treatment, or it may purchase reclaimed water for certain uses. Industrial water reuse has increased substantially because industrial water requirements have grown significantly (California Department of Water Resources 1982). The most promising avenue by which industrial demands may be reduced is increased reuse or multiple use within the plant. This usually entails treatment of the plant effluent so that the water quality meets effluent requirements and can be returned as part of the intake (Kollar and Brewer 1980b).

In industrial applications, the gross demand for water is met by a combination of once-through water use and the recycling of treated effluents. Once-through water use is the conventional method of using fresh water for cleaning and other purposes. Water not incorporated into the product is usually discharged after a single use. Once-through water use produces large quantities of low-strength wastewater, which must be treated and discharged. Although in-plant conservation can reduce the costs of water intake and waste discharge by as much as 50 percent, once-through use entails higher costs for wastewater treatment and water intake than can be obtained when a recycling system is used (California Department of Water Resources 1982).

In a reuse system, process water may be recycled continuously, with fresh water added only to compensate for evaporation losses, to supply product filling, and to maintain the required water quality. An example of industrial water recycling is the continuous use of water in an industrial cooling tower. Although the water is continuously recycled and reused, additional fresh water must be introduced, because some of the water evaporates, some of it becomes excessively mineralized and must be discharged as blowdown, and some of it escapes as drift (droplets of water suspended in the air leaving the tower). All of these losses (evaporation, blowdown, and drift) must then be replaced with additional water (called *makeup*) of lower mineral concentration.

Kollar and Brewer (1980*b*) describe the gross water use required in any manufacturing process as directly related to production; the quantity of intake water supplied depends on the recirculation rate, which in turn is influenced by the following factors:

- Availability and cost of water delivered to the plant
- Quality characteristics of the raw water
- Plant processes and plant technology
- Recovery of materials, products, by-products, and energy
- Consumption losses
- Air and water pollution control regulations
- Cost avoidance
- Age of plant

The most important of these factors is the (increasingly rigorous) air- and water-pollution control regulations.

A bulletin published by the California Department of Water Resources (1982) included calculations of recirculation rates (gross freshwater use divided by intake) for major industry groups in the SIC system (table 1-3). Major industries that reported high recycling rates included paper mills, petroleum refining, rubber and plastic products, primary metals, and electrical and electronic equipment.

**Table 1-3. Recirculation Rate, by Industry Group.**

| SIC Code | Industry Group | Recycle Rate (percentages) |
|---|---|---|
| 20 | Food and kindred products | 3 |
| 21 | Tobacco manufacturers | 0 |
| 22 | Textile mill products | 18 |
| 23 | Apparel and other textile products | 1 |
| 24 | Lumber and wood products | 2 |
| 25 | Furniture and fixtures | 2 |
| 26 | Paper and allied products | 9 |
| 27 | Printing and publishing | 2 |
| 28 | Chemicals and allied products | 6 |
| 29 | Petroleum and coal products | 19 |
| 30 | Rubber and plastic products | 17 |
| 31 | Leather and leather products | 3 |
| 32 | Stone, clay, glass, and concrete products | 5 |
| 33 | Primary metal industries | 10 |
| 34 | Fabricated metal products | 7 |
| 35 | Machinery, except electrical | 3 |
| 36 | Electrical and electronic equipment | 9 |
| 37 | Transportation equipment | 5 |
| 38 | Instruments and related products | 11 |
| 39 | Miscellaneous manufacturing industries | 11 |

SOURCE: California Department of Water Resources (1982), pp. 51–54.

## Factors Affecting Industrial Water Requirements

Manufacturing-plant water intake depends on such factors as the quality and type of raw material involved, the design of the plant, and the efficiency of the industrial processes used (California Department of Water Resources 1982). For those reasons, two plants manufacturing identical products may require substantially different quantities of water per unit of product.

Industrial water demands can be reduced significantly by recycling;

in some industries, however—particularly the food industry—reuse systems may conflict with sanitation requirements, except for such operations as container cooling, floor washing, and other cleaning, where the recycled water does not come into direct contact with the product. Each industry has its own particular water-quality requirements, which relate mainly to the interaction between certain chemicals or other constituents and the product. Each industry also has different potential applications for reclaimed water. Water-quality requirements (and reclaimed water potential) for each of the SIC system's major industry groups are described in the following paragraphs.

**Food Processing (SIC Major Group 20)**   In the food industry, potentially harmful contaminants and characteristics include suspended solids, heavy metals, turbidity, and coliform bacteria. Reclaimed water can be used for floor- and gutter-washing, boiler-feed water, and initial raw-product conveyance. Reclaimed water is unacceptable, however, for certain processes in which water comes into direct contact with the product. In the meat-processing industry, for example, reclaimed water can be used for cleanup and sanitation only.

**Textiles (SIC Major Group 22)**   Of concern in the textile industry are calcium and magnesium, which can interact with soaps during the washing process, cause dyes to precipitate, and produce color irregularities between batches of dyed material.

**Paper Products (SIC Major Group 26)**   Paper manufacturers are concerned with color, turbidity, and suspended solids. Water hardness is also a problem, because certain mineral constituents of water may react with sizing compounds; in addition, organisms in reclaimed water may cause slime and other growths on the paper. Manufacturers of lower-grade papers can more easily use reclaimed water for processing. In general, the finer the grade of paper, the more critical water quality becomes.

**Chemical Industry (SIC Major Group 28)**   Processing in the chemical industry is highly specific, and each industry has its own special requirements. In general, manufacturers of inorganic chemicals can use reclaimed water more easily than can manufacturers of organic chemicals.

**Petroleum Refining (SIC Major Group 29)**  In the petroleum industry, some 75 to 85 percent of the water withdrawn is used for cooling. Troublesome constituents of cooling water include calcium (hardness), phosphates, silica, and ammonia. In boiler-feed water, the main complicating factors are hardness and total dissolved solids (TDS). Of concern in process water are TDS, bacteria, and suspended solids; in general, using reclaimed water for processes would conserve large quantities of fresh water but would involve high costs for pretreatment and treatment of effluents.

**Glass and Glassware (SIC Major Group 32)**  In glass manufacturing, the use of reclaimed water poses few problems, since water neither enters the product nor comes into contact with it.

**Primary Metals (SIC Major Group 33)**  Water-quality requirements for the primary-metals industry vary, depending on whether the water is used for quenching, hot-rolling, cold-rolling, or rinsing. Regardless of the process, however, most metals manufacturers could use reclaimed water to their advantage.

### Forecast Guidelines

In forecasting industrial demands for a single industrial plant, the water planner usually calculates unit water demands on the basis of operating experience at similar plants, readings from pilot operations, or both (Kollar and Brewer 1980b). For most industrial establishments, information is available on water demand per unit of production for the technology to be used. When forecasting future water demands for a major metropolitan area, however, the water planner may be dealing with forecasts of demands for several thousands of manufacturing establishments.

During the planning period, typically 30 to 50 years, the manufacturing sector may increase in size and output by 500 to 700 percent. An additional complicating factor is that projections of growth for the manufacturing sector are not expressed in terms of material production—tons of steel, for example, or barrels of oil—but according to such parameters as employment, or dollars of earnings. Moreover, these parameters are not reported industry-by-industry, but for the manufacturing sector as a whole or, at best, for groups of similar industries.

Until the early 1960s, the common way of projecting industrial water

**Table 1-4. Unit Use per Employee by Industry Group.**

| SIC Code | Industry Group | Unit Use/ Work Day (gallons) |
|---|---|---|
| 20 | Food and kindred products | 2,026 |
| 21 | Tobacco manufacturers | 100 |
| 22 | Textile mill products | 43 |
| 23 | Apparel and other textile products | 38 |
| 24 | Lumber and wood products | 2,253 |
| 25 | Furniture and fixtures | 104 |
| 26 | Paper and allied products | 3,454 |
| 27 | Printing and publishing | 81 |
| 28 | Chemicals and allied products | 1,443 |
| 29 | Petroleum and coal products | 9,392 |
| 30 | Rubber and plastic products | 290 |
| 31 | Leather and leather products | 169 |
| 32 | Stone, clay, glass, and concrete products | 1,146 |
| 33 | Primary metal industries | 772 |
| 34 | Fabricated metal products | 315 |
| 35 | Machinery, except electrical | 158 |
| 36 | Electrical and electronic equipment | 162 |
| 37 | Transportation equipment | 173 |
| 38 | Instruments and related products | 213 |
| 39 | Miscellaneous manufacturing industries | 61 |

SOURCE: California Department of Water Resources (1982), pp. 76–80.

demands was to calculate industrial water needs on the basis of general population growth. As data on water use improved and statistics on industrial employment became available, water use per employee became the usual parameter for projections. Unit water-use rates per employee were estimated in the California Department of Water Resources survey (1982) and are summarized in table 1-4. Major industries reporting high rates of water use per employee include those producing petroleum and coal products, lumber and wood products, food and food-related products, and chemicals and related products.

Projecting industrial demands on the basis of rates of water use per employee is susceptible to error. Industrial development rarely occurs at the same rate as population growth in a developing economy.

Although changes in manufacturing employment may be good indicators of the distribution of income, they fail as indicators of water use because they lack a correlation to employee productivity or to the kinds and types of industries that may locate in a study area.

Kollar and Brewer (1980*b*) propose using the physical output of the industry sector as the parameter of choice for projecting industrial water demands. If this information is not available, the next best parameter is a proxy measure of production, such as value added by manufacture or gross product originating (GPO), for which there are some historical data. Value added by manufacture is the difference in dollars between the value of the industry's shipments and the costs of material, energy, labor, and overhead. Gross product originating is the portion of gross national product (GNP) attributed to the industry group under study. Both are expressed in dollar terms. These figures can then be corrected for changes in real value as a result of inflation or deflation, to enable the dollar unit to maintain a constant value. Economists are now furnishing such units as they adjust their thinking to water-resources planning.

## COMMERCIAL DEMANDS

In their study of commercial water demand, Wolff, Linaweaver, and Geyer (1975) derived water-use factors for commercial establishments from data collected in the early 1960s. Twenty-nine separate commercial and institutional facilities in the Baltimore metropolitan area were monitored for periods of at least three years each. Water use by each facility was evaluated from quarterly billing records, from recorders installed on the water meters, or from visual meter readings at daily and hourly intervals. The data were used to develop design criteria for water-distribution systems and to establish equitable water-con-nection charges and sewer-service charges.

A subsequent study, by McCuen, Sutherland, and Kim (1980), focused on water use by department stores and other commercial establishments common in urban and suburban shopping centers. The study used data from four shopping centers (one in Greendale, Wisconsin; one in Baltimore, Maryland; and two in suburban Washington, D.C.), comprising two department stores with restaurants, five department stores without restaurants, and 140 mall shops of different types.

The McCuen study offered two explanations for why water demand by the commercial sector has not been a primary factor in the design of urban water systems. First, commercial water use constitutes only a small proportion of total water demand in a typical urban water system. Commercial water use normally accounts for only 15 to 20 percent of the total demand, whereas residential and industrial water uses may account for more than 70 percent. Second, commercial water use is concentrated at a favored location in the demand hydrograph. According to the Wolff study, commercial demands during peak hour typically represent only about 4.4 percent of the total peak demand. The largest component of demand during peak hour is for residential lawn sprinkling.

Commercial water demands do, however, play a large role in urban water systems. The United States' economy is currently in a period of transition from an industrial-oriented economy to a service-oriented economy; more than 55 percent of the total employment is now in the service sector (McCuen, Sutherland, and Kim 1980). In addition, a closer inspection of commercial water demand is warranted in many situations. In areas experiencing rapid growth, for example, commercial establishments often run afoul of sewer-hookup moratoriums imposed to prevent sewage-treatment facilities from becoming overloaded. Individuals proposing commercial development must provide forecasts of commercial water use. Since past investigations have been concerned primarily with the residential and industrial sectors, a reliable method of forecasting water use by commercial establishments has not yet been devised.

## Types of Commercial Demand

Commercial and industrial businesses in the United States are defined under division G, "Retail Trade," and division I, "Services," in the SIC system (table 1-5). Some commercial establishments may be found in other major subgroups.

Table 1-6 summarizes the results of the study by Wolff, Linaweaver, and Geyer, showing annual, maximum-day, and peak-hour water use for selected commercial establishments. For each type of commercial establishment, water use is expressed in terms of a parameter that best describes the business's water-use characteristics. The difference between "expected" and "design" water-use factors in table 1-6 reflects the variability in water-using habits among facilities of the same type. Design values are set sufficiently high to ensure that actual use in any one facility will not exceed the design value. For some types of facilities, the number of samples was not sufficient to allow selection of a design value.

**Table 1-5. Commercial Classifications.**

| Major Group Number | | Classification | Major Group Number | | Classification |
|---|---|---|---|---|---|
| Division G | 52 | Building materials, hardware, garden supply, and mobile home dealers | Division I | 75 | Automotive repair, services, and garages |
| | 53 | General merchandise stores | | 76 | Miscellaneous repair services |
| | 54 | Food stores | | 78 | Motion pictures |
| | 55 | Automotive dealers and gasoline service stations | | 79 | Amusement and recreation services, except motion pictures |
| | 56 | Apparel and accessory stores | | 80 | Health services |
| | 57 | Furniture, home furnishings, and equipment stores | | 81 | Legal services |
| | | | | 82 | Educational services |
| | | | | 83 | Social services |
| | 58 | Eating and drinking places | | 84 | Museums, art galleries, botanical and zoological gardens |
| | 59 | Miscellaneous retail | | 86 | Membership organizations |
| Division I | 70 | Hotels, rooming houses, camps, and other lodging places | | 88 | Private households |
| | 72 | Personnel services | | 89 | Miscellaneous services |
| | 73 | Business services | | | |

SOURCE: Office of Statistical Standards (1972).

During the last two decades, regional shopping centers have accounted for an increasing percentage of total commercial sales volume. The study by McCuen, Sutherland, and Kim offers an insight into the water-use characteristics of department stores and other commercial establishments (mall shops) usually found in regional shopping centers. It concluded that gross store area can be used as a proxy variable for water consumption in department stores and mall shops because it is a well-defined factor on which data are readily available. Quantitative estimates of other variables—such as total sales volume, sales tax revenue, sales area, and average man-hours of employment—are difficult to obtain. For example, department stores do not generally release sales volume figures or data on average man-hours of employment, and states do not release information about sales tax receipts for individual stores.

**Table 1-6. Summary of Commercial and Institutional Water Use.**

| Type of Establishment or Institution | Selected Parameter | Annual Water Use (gpd/unit) | | Maximum Day Water Use (gpd/unit) | | Peak Hour Water Use (gpd/unit) | |
|---|---|---|---|---|---|---|---|
| | | Expected | Design | Expected | Design | Expected | Design |
| Primary and secondary schools | | | | | | | |
| public elementary | gpd/student | 5.38 | 8.67 | 9.68 | 13.00 | 49.10 | 52.40 |
| public senior high | gpd/student | 5.64 | 9.75 | 19.60 | 25.20 | 121.00 | 127.00 |
| public junior high | gpd/student | 6.63 | 12.20 | | | | |
| private elementary | gpd/student | 2.24 | 6.09 | 3.10 | 6.92 | 25.70 | 29.50 |
| private senior high | gpd/student | 10.40 | 18.60 | 15.70 | 23.90 | 38.70 | 46.90 |
| combined (grades 1–12) | gpd/student | 8.49 | 18.70 | 16.80 | 27.00 | 51.30 | 61.50 |
| Colleges | | | | | | | |
| students in residence | gpd/student | 106 | 179 | 114 | 187 | 250 | 323 |
| nonresident students | gpd/student | 15 | * | 27 | * | 58 | * |
| Hospitals | gpd/bed | 346 | 559 | 551 | 764 | 912 | 1120 |
| Nursing homes and institutions | gpd/bed | 113 | 209 | 146 | 222 | 424 | 500 |
| Apartments | | | | | | | |
| high-rise | gpd/occupied unit | 218 | 322 | 426 | 530 | 745 | 849 |
| garden-type | gpd/occupied unit | 213 | 315 | 272 | 374 | 671 | 773 |

| | | | | | | |
|---|---|---|---|---|---|---|
| Hotels | gpd/sq. ft. | 0.256 | * | 0.294 | * | 0.433 | * |
| Motels | gpd/sq. ft. | 0.224 | 0.326 | 0.461 | 0.563 | 1.55 | 1.65 |
| Office buildings | | | | | | | |
| general offices (− 10 years) | gpd/sq. ft. | 0.093 | 0.164 | 0.173 | 0.244 | 0.521 | 0.592 |
| general offices (+ 10 years) | gpd/sq. ft. | 0.142 | 0.273 | | | | |
| medical offices | gpd/sq. ft. | 0.618 | * | 1.660 | * | 4.970 | * |
| Department stores | gpd/sq. ft. of total sales area | 0.216 | 0.483 | 0.388 | 0.655 | 0.958 | 1.230 |
| Shopping centers | gpd/sq. ft. of total sales area | 0.160 | * | 0.232 | * | 0.412 | * |
| Car washes | gpd/sq. ft. | 4.78 | * | 10.3 | * | 31.5 | * |
| Service stations | gpd/sq. ft. of garage & office space | 0.251 | 0.485 | 0.590 | 0.824 | 4.890 | 5.120 |
| Laundries | | | | | | | |
| commercial laundries & dry cleaners | gpd/sq. ft. | 0.253 | 0.639 | 0.326 | 0.712 | 1.570 | 1.960 |
| laundromats | gpd/sq. ft. | 2.170 | 6.390 | * | * | 4.890 | * |
| Restaurants | | | | | | | |
| drive-ins (parking only) | gpd/car space | 109.0 | | | | | |
| drive-ins (seating & parking) | gpd/seat | 40.6 | | | | | |
| conventional restaurants | gpd/seat | 24.2 | 55.2 | 83.4 | 114 | 167 | 198 |

(continued)

# Table 1-6. (continued)

| Type of Establishment or Institution | Selected Parameter | Annual Water Use (gpd/unit) | | Maximum Day Water Use (gpd/unit) | | Peak Hour Water Use (gpd/unit) | |
|---|---|---|---|---|---|---|---|
| | | Expected | Design | Expected | Design | Expected | Design |
| Clubs | | | | | | | |
| golf | gpd/membership | 66.1 | | | | | |
| swimming | gpd/membership | 16.5 | | | | | |
| boating | gpd/membership | 10.5 | | | | | |
| Churches | gpd/member | 0.138 | | 0.862 | | 4.700 | |
| Barber shops | gpd/chair | 54.6 | 97.5 | 80.3 | 123 | 389 | 432 |
| Beauty salons | gpd/station | 269 | 532 | 328 | 591 | 1070 | 4330 |

* For some establishments, the number of samples was not sufficient to enable the selection of a design value.

SOURCE: Adapted from Wolff, Linaweaver, and Geyer (1975), p. 49, by permission of ASCE.

Water use by department stores was calculated differently depending on whether or not they contained restaurants, because water use increases significantly when a restaurant is present. The water use at the five department stores without restaurants ranged from 1,813 gpd (gallons per day) to 10,831 gpd, with a mean of 6,555 gpd and a standard deviation of 3,262 gpd. The department stores ranged in gross area from 88,000 square feet to 218,800 square feet, with a mean of 145,760 square feet and a standard deviation of 53,660 square feet. The mean water-use factor was 0.04348 gallons per working day per square foot of gross area (gdsf). Water use at the two department stores containing restaurants was 9,206 gpd and 15,108 gpd. The gross areas of the two stores were 159,000 square feet and 179,000 square feet, resulting in water-use factors of 0.0579 gdsf and 0.0844 gdsf.

Other conclusions of the McCuen study can be summarized as follows:

**1.** A close correlation exists between sales area and (the more easily measured) gross store area.
**2.** The number of restrooms in a store correlates closely with gross store area.
**3.** A good correlation seems to exist between the number of employees and the number of customers, for department stores having similar economic functions.
**4.** Since facilities involving water use in department stores are available to customers as well as to employees, estimates of both employee and customer use of water are required.
**5.** The size of the region that the department store serves, the population of the region, and the economic class of the clientele may also influence water use in department stores.

McCuen, Sutherland, and Kim found that water use in mall shops was considerably less than water use in department stores, because the restroom facilities in mall shops are usually not available for customer use. Mall shop employees use water primarily in the restrooms, with smaller amounts for making coffee, for cleaning, and so on. In some specialty shops, water is also used for cleaning merchandise— such as wigs in a wig shop—or for in-house film development in some camera shops. Thus, the type of mall shop influences water use; and therefore, water-use relationships should be derived for each mall shop classification.

   In addition to finding variations in water use among the different types of mall shops, the McCuen study found variations in water use within each classification. Since most water use results from personal use by employees, the number of employees undoubtedly influences the total water use. Irregularity in the length of employee work shifts, however, could be responsible for enough variation in water use to make the number of employees, by itself, an inadequate indicator of water use. A comparison of water-use factors for mall shops was derived by McCuen, Sutherland, and Kim and is shown in table 1-7.

## Forecasting Applications

The methodology of long-range water-demand forecasting for commercial establishments is similar to the methodology currently used for residential and industrial water-demand forecasting (McCuen, Sutherland, and Kim 1980). Factors that could influence residential demand (regulations, pricing policy, educational campaigns, housing trends, supply costs, and change in technology of demand) could also influence commercial demand. But the importance of these variables in commercial water-demand forecasting is different from their importance in residential water-demand forecasting.

   Building code regulations could be responsible for even greater savings in commercial establishments than in residential developments because a large percentage of the total commercial demand is caused by toilet-flushing and lavatory use. The use of water-saving devices such as spring-controlled faucets (which would not even be considered in residential bathrooms) could offer significant savings in water use.

   Pricing policies are probably less important in forecasting the water demands of commercial establishments than they might be in forecasting those of residential developments because the water user is not directly responsible for water costs. An analogy can be made between the commercial water user and a renter in a multifamily residential development, from whom the true cost of water is concealed within the monthly rent. An educational campaign for water conservation in the workplace would not be an important factor in commercial water forecasting, since the largest portion of total water demand is for domestic-type uses. The owner of a store might make the employees aware of the impact of water costs on overall store costs, but the resulting water savings (if any) probably would not significantly affect total urban water demands.

**Table 1-7. Water-Use Factors.**

| SIC | Function | N | Mean Water Use (gpd) | Std. Dev. Water Use (gpd) | Range of Gross Store Area (sq. ft.) | Mean Water Use (gdsf × 100) | Std. Dev. Water Use (gdsf × 100) |
|-----|----------|---|----------------------|---------------------------|--------------------------------------|------------------------------|-----------------------------------|
| 5499 | Health food | 1 | 35.73 | — | 1,222 | 2.92 | — |
| 5611 | Men's clothing | 20 | 127.68 | 132.34 | 930–17,950 | 2.23 | 0.79 |
| 5621 | Women's clothing | 32 | 76.09 | 53.92 | 796–10,594 | 2.58 | 1.46 |
| 5631 | Hosiery | 1 | 154.57 | — | 1,225 | 12.62 | — |
| 5661 | Shoes | 28 | 67.79 | 36.50 | 6,000–63,000 | 2.74 | 1.60 |
| 5699 | Wigs | 2 | 65.05 | 15.74 | 11,000–27,500 | 8.49 | 0.91 |
| 5699 | Uniforms | 2 | 105.68 | 95.00 | 750–1,522 | 8.22 | 4.37 |
| 5713 | Carpets | 2 | 38.14 | 4.18 | 1,125–3,023 | 2.40 | 1.76 |
| 5719 | Cutlery | 1 | 26.87 | | 1,053 | 2.55 | |
| 5722 | Appliances | 4 | 163.54 | 213.74 | 2,009–26,652 | 1.76 | 0.15 |
| 5733 | Music | 4 | 109.87 | 61.10 | 11,000–40,000 | 3.79 | 0.61 |
| 5941 | Sporting goods | 1 | 57.62 | — | 5,338 | 1.08 | — |
| 5942 | Books | 3 | 50.92 | 9.53 | 3,437–6,480 | 1.17 | 0.43 |
| 5944 | Jewelry | 7 | 89.89 | 37.99 | 12,200–38,000 | 5.05 | 2.60 |
| 5945 | Toys | 3 | 61.51 | 27.63 | 1,141–5,269 | 2.10 | 0.57 |
| 5946 | Cameras and photography | 2 | 51.90 | 21.73 | 720–1,411 | 4.89 | 0.25 |
| 5947 | Gifts and stationery | 12 | 48.25 | 19.95 | 5,800–25,800 | 2.16 | 0.89 |
| 5949 | Fabric | 7 | 117.28 | 47.86 | 20,400–70,000 | 1.78 | 0.36 |
| 5999 | Art supplies | 2 | 38.92 | 16.64 | 1,440–3,149 | 1.75 | 0.20 |
| 5999 | Cosmetics | 1 | 34.15 | — | 520 | 6.57 | — |
| 7999 | Art gallery | 1 | 28.53 | — | 3,776 | 0.76 | — |
| 5399 | Bath goods | 1 | 63.83 | — | 3,820 | 1.67 | — |
| 5499 | Gourmet foods | 1 | 147.32 | — | 3,699 | 3.98 | — |
| 8091 | Opticians | 2 | 40.64 | 18.91 | 760–1,318 | 3.84 | 0.36 |
| | Total | 140 | 82.69 | | | 2.86 | |

SOURCE: Reprinted from *Energy and Water Use Forecasting*, by permission; copyright © 1980, American Water Works Association.

Housing trends must be considered in commercial water-demand forecasting. The transition from a neighborhood-oriented commercial economy to a region-oriented commercial economy may increase water use by commercial establishments because of the longer time spent on a single shopping trip to a regional shopping center. Thus, water used at the shopping center for domestic-type purposes substitutes for a certain amount of residential water use. If the trend toward regional shopping centers continues, greater water use by commercial establishments may result.

The impact on commercial establishments of technological advances in water supply systems—such as self-reading meters or less-expensive plastic pipe—will probably resemble the impact of the same advances on residential developments. On the other hand, the effect of technological changes on water demand may be greater in commercial establishments than in the residential sector. To increase the attractiveness of a shopping center (and thereby to attract a larger share of the market), commercial establishments may increase their use of water-based displays (such as fountains), or they may undertake extensive landscaping. In a typical commercial establishment, domestic-type water use and air conditioning account for most of the total water demand. Subpotable water might be enlisted for some of these uses.

McCuen, Sutherland, and Kim (1980) suggest that the following additional variables be examined in commercial water-demand forecasting: layout and design characteristics, transportation balance, and labor productivity. Modifications of current layout and design characteristics could help control employee and consumer water use. For example, at shopping centers, restroom facilities for employees are provided at each mall shop; considerable reductions in water use, construction costs, and lost floor space could be achieved by having employees use centrally located restrooms.

Commercial water-demand forecasting models should reflect changes in the transportation balance. Although automobile travel is expected to increase, many planners expect the current transportation balance to be altered by modern mass-transit systems, which may reverse the current trend and increase the percentage of commercial activity in the central business district. The slight inconvenience of mass-transit systems may increase average shopping-trip lengths and consequently may increase water use at commercial establishments.

Because commercial establishments are service-oriented, they are characterized by a comparatively low output/labor ratio. Changes in employee productivity directly influence water demand. Many service-oriented enterprises, such as gasoline stations that have self-service pumps, are becoming less labor-intensive. Continuation of this trend would cause a decrease in water use at commercial establishments.

## UNACCOUNTED-FOR WATER USE

After examining the major components (residential, industrial, and commercial) of urban water demand, the water planner can focus on unaccounted-for water (UAW)—a component of urban water demand that, in the past, has been regarded as analytically insignificant. The recent appearance of comprehensive studies of UAW, however, indicates a change in management philosophy toward this water-use component. *Unaccounted-for water use* is defined as the difference between a water utility's production and its water sales to consumers. Thus, total water produced minus water sold equals UAW.

### Authorized versus Unauthorized UAW

A distinction should be made between authorized UAW uses and unauthorized UAW uses. Authorized uses are specific unmetered uses of water that the water utility recognizes to be beneficial and/or necessary. One or more of the following uses may be considered authorized UAW uses by the water utility:

1. Unmetered use from fire hydrants
   - Fire-fighting and training
   - Street-washing
   - Construction
   - Main-flushing
   - Sewer-cleaning
   - Storm drain—flushing
2. Unmetered connections
   - Public buildings
   - Schools
   - Golf courses
   - Cemeteries
   - Parks
3. Reservoir seepage and evaporation

Unauthorized UAW losses comprise all other uses, losses, and measuring errors. Examples of unauthorized UAW losses include the following:

1. Water system leakage
2. Unauthorized use
   • Theft
   • Deliberate bypass of meters
   • Illegal tapping
   • Abandoned services
3. Inaccurate meters
4. Inadequate system controls
   • Malfunctioning valves
   • Overflowing reservoirs
5. Incorrect meter-reading and billing

Few, if any, water systems are entirely free of UAW losses of both kinds. Given normal inaccuracies in water meters, meter readings, and normal water system design and operations, a water system can be expected to operate with at least 2 percent UAW. A survey conducted for the California Department of Water Resources' Office of Water Conservation (1982), estimated total UAW at 8.4 percent on a volumetric basis, with typical utilities having total UAW loss levels of 9.3 percent.* For long-range planning, statewide UAW is estimated to be 9.5 percent of municipal water production.

Typically, unauthorized uses far exceed authorized uses of UAW. In the Office of Water Conservation survey, authorized UAW use constituted 1.5 percent of total water production on a volumetric basis; unauthorized UAW losses were, therefore, the difference between the total UAW (9.5 percent) and the authorized UAW use (1.5 percent), or 8.0 percent of water production. A survey conducted by the American Water Works Association used a utility's UAW percentage as a measure of its system's performance (Moyer 1985). Systems having 10 to 15 percent UAW are considered to be performing very well, and system losses of up to 20 percent are considered reasonable. In an unusual application of UAW use figures, a Lake Michigan water-supply con-

---

* Percent UAW is expressed as a percentage of total water produced. When percent UAW data from many water utilities are summarized, the average UAW can be expressed either as a typical water utility percentage or as a volumetric percentage. The typical water utility percentage is the average percentage of UAW for all water utilities, regardless of their size, while the volumetric percent UAW is a weighted average giving added weight to larger water utilities.

Table 1-8. California Urban Water Production and Losses.

| Type | Percent | Quantity (acre-feet) |
|---|---|---|
| Total water produced | 100 | 6,100,000 |
| UAW | 9.5 | 579,500 |
| Authorized UAW | 1.5 | 91,500 |
| Unauthorized UAW | 8.0 | 488,000 |
| meter error | 4.0 | 244,000 |
| leakage | 4.0 | 244,000 |
| Economically recoverable leakage | 1.0–1.5 | 61,000–91,500 |

SOURCE: California Department of Water Resources (1982).

sortium required each of its member agencies to show a UAW loss of less than 8 percent in order to secure bonding for a new pipeline project (*American City & Country* 1986).

In 1985, Philadelphia's water department determined that as much as 148 million gallons of water per day were being lost from illegally opened fire hydrants, costing the city almost $500,000 a year (Fredette 1986). A public-relations campaign aimed at children from eight to twelve years old—those most likely to play at open hydrants—reduced the amount of water lost by 46 percent.

## Types of Unauthorized UAW

By far the largest sources of unauthorized UAW losses are believed to be metering errors and leakage; other sources have a negligible effect. The Office of Water Conservation survey concluded that metering errors were responsible for between 40 and 60 percent of unauthorized UAW losses, with the chief culprits being older meters and under-registration at low flows. Statewide in California, the average share of unauthorized UAW losses attributable to metering errors is estimated to be 50 percent—that is, 4 percent of total water production. The remaining 50 percent of unauthorized UAW losses are thought to be caused primarily by leakage in water distribution systems, from leaks in water mains, valves, fittings, hydrants, and at meters. Table 1-8 summarizes the percentage breakdown of UAW.

In the American Water Works Association survey, leakage was estimated to account for 39 to 70 percent of UAW, and undermetering was estimated to account for 18 to 48 percent (Moyer 1985). Undermetering includes both meter underregistration and water provided to unmetered services. In view of these estimates, combating problems of leakage and undermetering should be given high priority. The other components of UAW (authorized unmetered public services and unauthorized use) are generally minor; estimates of these components range from 1.0 to 7.2 percent of total water production (that is, 5 to 18 percent of UAW).

**Meter Error**   Because meters determine revenue, the amount of revenue a water utility receives depends on the accuracy of its meters. Meters can overregister or underregister, but the net percentage of error directly reflects the net percentage of distribution system losses. It is important that meters be of the proper size and type. An excessively large meter may not register low flows; an undersized meter may produce extreme drops in pressure that cause the meter to wear out in a short time. Provisions should be made for testing meters already in place, preferably without interrupting service.

Meters can be divided into three general categories: master meters, industrial and commercial meters, and domestic meters. Master meters are located at the supply sources that deliver water to the distribution system. Among the different types of master meters are venturi, orifice plate, and impeller varieties; in some installations, turbines or compound meters can be used. All master meters should be tested at initial installation and at least once per year thereafter. Enough straight pipe should be provided ahead of the meter at installation to ensure maximum uniformity of flow and (consequently) accurate registration.

In many cities, industrial and commercial meters register a large percentage of water utility revenue. The frequency with which industrial meters should be tested depends on their flow rate. Meters supplying large-volume users of water should be tested at least once a year, and more often if the revenue involved is large. All 3-inch or larger meters should have test plugs or other provisions that allow for testing them in place without interrupting service. Accurate portable test meters should be available for field tests, and manufacturer's recommendations should be followed in installing these meters.

Domestic meters are the principal revenue registers in many suburban cities and are important (although somewhat less so) in cities where a large amount of revenue is registered on industrial and commercial meters. In spite of this fact, domestic meters are the most neglected part of the water system. A domestic meter can run for many years

**Table 1-9. Percentage of Total Flow through Domestic Meters at Various Rates.**

| Rate of Flow | | Water Used |
|---|---|---|
| l/s | gpm | (percent of total) |
| 0–0.015 | 0–0.25 | 13.0 |
| 0.015–0.03 | 0.25–0.50 | 3.4 |
| 0.03–0.06 | 0.50–1 | 6.8 |
| 0.06–0.12 | 1–2 | 13.3 |
| 0.12–0.24 | 2–4 | 43.0 |
| more than 0.24 | more than 4 | 20.5 |

SOURCE: Reprinted from *Water Conservation Strategies*, by permission; copyright © 1980, American Water Works Association.

before it stops completely, and for most of its operating life it under-registers and thus loses revenue for the water utility. As soon as any meter is installed and begins to operate, its internal parts begin to wear, causing underregistration at low flows. The longer the wearing process continues, the greater the degree of underregistration, with the rate of wear depending on the properties of the water.

The amount of water used in a water system at low rates of flow is larger than is commonly realized. Hudson (1980) combined the results of several studies to determine the amount of water used at various rates of flow through domestic meters. The data, summarized in table 1-9, show that approximately 23 percent of all water for domestic use had been withdrawn at flow rates of 1 gpm (0.06 liter per second) or less. This is the range at which domestic meters first start to underregister.

Hudson indicates that, ideally, domestic meters should not be left in service more than eight to ten years before they are rebuilt; Hudson's analysis is summarized in table 1-10. In water that is relatively corrosive or that contains sand and grit, the service period may be shorter. At least 20 percent of all meters with more than eight or ten years of service fail to register flows of less than 0.75 gpm (0.045 l/s). Since 23 percent of all domestic water is used at 1 gpm (0.06 l/s) or a lower flow rate, approximately 5 percent of a utility's revenue can be lost as a result of domestic meters that fail to register low flows.

Underregistering water flow amounts to giving away water; and if sewer rental charges are based on metered water consumption, the

**Table 1-10. Underregistration of Meters According to Age.**[1]

| Age of Meters (years) | Inaccurate Meters (percent) | Minimum Registrable Flow | |
|---|---|---|---|
| | | l/s | gpm |
| 0–9 | 5 | 0.045 | 0.75 |
| 9–19 | 20 | 0.045 | 0.75 |
| 19–29 | 50 | 0.075 | 1.25 |
| more than 29 | 84 | 0.09 | 1.5 |

[1] Averages based on four separate studies.

SOURCE: Reprinted from *Water Conservation Strategies*, by permission, copyright © 1980, American Water Works Association.

water utility loses still more revenue. Such losses can be reduced if all meters are read at regular intervals. Stopped meters and meters with unusual or out-of-line readings should be tested and repaired. Computer billing systems can be programmed to record and compute the metered consumption and flag unusual readings.

**Leakage** No water system is entirely free of leakage. Settling, expansion, and contraction of water mains and appurtenances, as well as industry standard tolerances for construction and testing, make a certain amount of water system leakage inevitable. Corrosion, poor construction practices, normal wear and tear, deterioration of material in gaskets, and accidents can contribute to more substantial water system leakage.

Leak-detection procedures cannot identify all sources of leakage, because of human and mechanical limitations. In areas with plentiful water at relatively low prices, a leak-detection program may even be economically unjustifiable. Rapidly escalating water and energy costs, however, indicate a need to recover UAW through successful leak-detection programs. Based on a review of successful leak-detection programs, California's Office of Water Conservation (1982) concluded that between 25 and 40 percent of the water lost through leakage in state water systems (1 to 1.5 percent of total production) was economically recoverable through a statewide program of leak detection.

Leak-detection programs may use various types of leak-detection equipment and techniques, entailing a wide range of costs. Leak-detection equipment is used for two distinctly different purposes: to search for sounds of leakage near the monitoring point; and to isolate and pinpoint the precise location of the leak sound, to enable a field maintenance crew to make the necessary repairs. Higher-quality leak-detection equipment can enable a utility to achieve 90 to 95 percent accuracy in exposing the leak within the initial excavation area. Techniques used to identify and locate municipal water-system leakage range from visual observation and isolated sector consumption studies to the use of sophisticated leak-detection equipment (oscilloscopes, microprocessors, accelerometers, and amplifiers). Basic acoustical equipment can be used to identify the sound of escaping water; more sophisticated equipment is used to attain greater degrees of accuracy in pinpointing leaks.

In isolated sector consumption studies, the distribution system is mapped into districts in such a way that water enters each district through a single main (feed main) when valves are closed on other mains leading into the district (Moyer 1985). When a district is surveyed, these other valves are closed; then the flow into the district through the feed main is continuously monitored by means of a Pitot tube and recorded for a 24-hour period. From the resulting record of flow rate versus time, the minimum and average rates of flow during the 24-hour period are determined, and their ratio is used as an indicator of the amount of leakage in the district. Since metered water consumption generally decreases at night, a high ratio of minimum water flow to average water flow implies a potentially high rate of leakage, while a low ratio suggests that leakage is probably insignificant.

Districts with high ratios of minimum flow to average flow are investigated further by means of nighttime flow measurements of district subdivisions covering several blocks, by means of sonic surveying (using geophones), or by means of visual inspection. The leakage rates of discovered leaks are measured with Pitot tubes. Following detection and repair of leaks in a district, water flow can be remeasured to see if additional serious leaks exist that were not detected the first time.

A benefit/cost analysis of a leak-detection and repair program compares the benefits derived, in terms of the value of water conserved, to the costs of leak detection and repair. In the American Water Works Association survey, the impact of leak detection and repair (LD&R) depends on the presence or absence of an unlimited supply of water for customers (Moyer 1985). In the first hypothetical situation (where water supply has been able to meet demand), avoided leakage results in a decrease in UAW, the amount of water sold is unaffected, and water production decreases. The savings to the utility resulting from LD&R are in the form of decreased production costs, since less water needs to be produced in order to meet demand. In the second situation (where supply has been unable to meet demand), avoided leakage results in a decrease in UAW, an increase in water sold to consumers, and a constant rate of water production. In this situation, the benefits to the water utility are in the form of increased revenues from water sales, since more of the produced water is sold.

A necessary first step in determining the economic feasibility of a leak-detection program is a water audit. Each utility's water system has individual characteristics that may need specific evaluation in order for the auditor to assess the actual amount of leakage. A water audit attempts to account for all water entering or leaving a distribution system and analyzes all components of UAW (unmetered uses, unauthorized uses, metering inaccuracies, and water system losses through leakage or operational malfunctions). The cost of a water audit can vary widely, depending on the utility's complexity and its record-keeping system.

Finally, the cost of a leak-detection program includes the manpower required to search for and repair leaks on a day-to-day basis. Leaks found in a leak-detection survey may vary from small drips that can easily be fixed with a wrench to major leaks requiring large expenditures for excavation and repair. The Office of Water Conservation survey found that, in California, approximately 60 percent of all leaks detected are minor, requiring only incidental spare parts and materials. The survey also reported that 20 to 40 percent of all leaks are found in the water main; the remaining 60 to 80 percent are found at the meter or in the customer's piping.

The American Water Works Association survey found that the mean leakage rates for different utilities generally fall within the same order of magnitude for a given leak category. With a few exceptions, main leaks had the highest leakage rates, followed by service leaks, hydrant leaks, and meter leaks. In addition, the survey found that large-diameter mains tended to develop joint leaks, whereas smaller-diameter main leaks tended to be caused by holes and breaks on the shaft of the pipe.

The relationship between leaks and breaks is central to the question of repair costs (Moyer 1985). Leaks can contribute to breaks by eroding the fill around a pipe, causing insufficient pipe support and leaving the pipe susceptible to damage from such forces as water hammer and traffic vibration. Breaks are generally more expensive to repair than are leaks, and they have a greater potential for damaging property and causing disruptions in other services. It has long been assumed (without proof) that leaks generally increase in size and severity over time. Therefore, most leaks are eventually detected and repaired. In addition to saving the cost of water leaked in the interim, early repair may be cheaper in terms of labor and replacement parts than later repair when the leak is further developed. For these reasons, the American Water Works Association survey suggests that repair costs not be included in the overall assessment of LD&R costs and benefits, since they probably do not add an expense greater than what the utility would normally pay for maintenance of the distribution system. For a crew that performs other functions besides leak detection, the costs of leak detection and repair can be prorated.

LD&R programs vary widely in how frequently they call for surveying the distribution system. Generally, smaller utilities are surveyed more frequently than larger utilities because of the time required to survey large water systems. The American Water Works Association reported that from one to fifteen years were required to complete one survey of an entire system. Large utilities that are committed to LD&R programs are surveying continually because it takes so long to do a survey of the entire system. Even though LD&R programs have yielded net benefits, little progress has been made in identifying the survey frequency at which net benefits are maximum. On one side of the maximum, survey costs are unnecessarily high because surveying is too frequent relative to the amount of leakage that is avoided as a result of repairs. On the other side of the maximum, the cost-effectiveness of LD&R is underutilized. The optimal surveying frequency would depend on the particular circumstances of the individual utility under consideration; it might also vary within the utility's service area for different regions, depending on leak and break propensity and on the potential for property damage.

It is often cost-effective to survey problem areas more frequently than other areas, particularly when the regular survey period is long. The following problem areas are typical (Moyer 1985):

- Areas with a history of excessive leak and break rates
- Heavily developed areas where leaks and breaks can result in severe property damage
- Areas exposed to stray electric current and traffic vibration
- Areas near stream crossings
- Areas where system pressure is high, since high pressure promotes leak formation
- Areas where loads on pipe exceed design loads

Benefits of a leak-detection program can be quantified in terms of the value of the water saved. A composite value for water conserved through leak detection can be developed by taking into account the following quantifiable benefits (Office of Water Conservation 1982):

1. *Cost of water.* The wholesale cost to purchase or produce water; it varies from district to district.
2. *Treatment cost.* The cost of chemicals and power to treat water.
3. *Energy cost.* The cost to distribute water throughout the system.
4. *Marginal cost of new facilities.* Conserving water can delay or eliminate the need for new sources, treatment, pumping, and storage facilities.

Other benefits from a leak-detection and repair program are not directly quantifiable in terms of dollars saved, but they are nonetheless real and advantageous to the utilities. Examples of nonquantifiable benefits include (Office of Water Conservation 1982):

1. *Public relations.* The rapidly rising cost of water makes it imperative that utilities demonstrate to the public their commitment to efficiency and conservation. If a utility does what it can to conserve water, customers will be more cooperative in other water-conservation programs—many of which require individual effort. An LD&R program can be highly visible, encouraging people to think about water conservation before they are asked to take action to reduce their own water use.
2. *Expertise gained/distribution-system control.* The knowledge gained about the condition and operation of valves, meters, and hydrants in the course of a leak-detection program can result in more efficient designs, material selection, and operations. For example, the information gained about the distribution system can be used in setting priorities for replacement or rehabilitation programs.

**3.** *Improved productivity.* The detailed knowledge gained of potential and existing problems can, through preventive maintenance or replacement, reduce emergency situations and thereby improve productivity.

**4.** *Damage prevention.* A leak-detection program can reduce the damage that leaking water mains cause to public and private property. The American Water Works Association survey reported that detection and repair of small leaks may prevent the development of costly breaks later on. Fewer main breaks and less property damage can lead to fewer claims and lower insurance rates.

Many California utilities embarked on leak-detection programs during the 1976–77 drought, but subsequently they reduced or ceased these operations. The most common reasons given for underutilized or unutilized leak-detection equipment were funding restrictions, reduction in UAW to acceptable levels, and lack of skilled operators. A lack of record-keeping or accounting of leak-detection benefits and costs precluded many utilities from assessing their programs' true costs and benefits.

## REFERENCES

*American City & County.* 1986. Streamwood streamlines water supply, revenues. *American City & County* (March 1986): 42.

American Water Works Association. 1986. Regionalization of water utilities. *MAINSTREAM* 30(4): 6–7.

Boland, J. J.; Moy, W. S.; Paccy, J. L.; Steiner, R. C. 1983. *Forecasting Municipal and Industrial Water Use: A Handbook of Methods.* U.S. Army Corps of Engineers, Engineer Institute for Water Resources.

Boyle Engineering Corporation. 1982. Municipal leak detection program, loss reduction—research and analysis. California Department of Water Resources, Office of Water Conservation.

California Department of Water Resources. 1982. Water use by manufacturing industries in California, 1979. CDWR Bulletin 124-3.

Coelen, S. P. 1981. Regional analysis in planning water supply extension. In *Selected Works in Water Supply, Water Conservation and Water Quality Planning,* ed. J. E. Crews and J. Tang, pp. 143–50. U.S. Army Corps of Engineers, Engineer Institute for Water Resources.

Fredette, J. B. 1986. Campaign halves hydrant losses. *American City & County* (March 1986): 46.

HUD. 1984. Residential water conservation projects—summary report. Report of the U.S. Department of Housing and Urban Development, Office of Policy Development and Research.

Hudson, W. D. 1980. Increasing water system efficiency through control of unaccounted-for water. In *Water Conservation Strategies*, pp. 94–97. Denver: AWWA.

Johnstone, G. W. 1985. Long-term planning for capital improvements. Paper presented at the AWWA Seminar on Demand Forecasting and Financial Risk Assessment, Washington, D.C., June 23, 1985.

Kollar, K. L., and Brewer, R. 1980*a*. The impact of industrial water use on public water supplies. In *Energy and Water Use Forecasting*, pp. 62–67. Denver: AWWA.

———. 1980*b*. Industrial development through water resources planning. In *Energy and Water Use Forecasting*, pp. 43–47. Denver: AWWA.

McCuen, R. H.; Sutherland, R. C.; and Kim, J. R. Forecasting urban water use: commercial establishments. In *Energy and Water Use Forecasting*, pp. 37–42. Denver: AWWA.

Miller, J. W., and Ludlum, M. D. 1985. Water demand forecasting and risk analysis (automating forecasting for risk management). Paper presented at the AWWA Seminar on Demand Forecasting and Financial Risk Assessment, Washington, D.C., June 23, 1985.

Moyer, E. E. 1985. *Economics of Leak Detection: A Case Study Approach.* Denver: AWWA.

National Water Commission. 1973. Water policies for the future. Final report of the National Water Commission. Washington, D.C.: Government Printing Office.

Office of Statistical Standards. 1972. *Standard Industrial Classification Manual.* Washington, D.C.: Executive Office of the President, Bureau of the Budget, Office of Statistical Standards.

Orange County Municipal Water District. 1985. Urban Water Management Plan (draft), July 1985.

Robinson, M. P., and Blair, E. R., Jr. Pump station design: the benefits of computer modeling. *Journal of the American Water Works Association* 76(7): 70–77.

Shelstad, M. J., and Hanson, H. 1986. Operations problems of small water systems. *Waterworld News* 2(1): 10–12.

Teknekron Research. 1979. Urban drought in the San Francisco Bay area: a study of institutional and social resiliency. Report prepared for the National Science Foundation, December 1979.

Topping, R. E., and Puccia, R. J. 1984. A computer-assisted water system analysis and design for Charleston, S.C. *Journal of the American Water Works Association* 76(7): 82–89.

U.S. Water News. 1986. More than water is being poured into irrigation. *U.S. Water News* 2(10): 16.

Wolff, J. B.; Linaweaver, F. P.; and Geyer, J. C. 1975. Water use in selected commercial and institutional establishments in the Baltimore metropolitan area. Report prepared by ASCE for the Office of Water Research and Technology, U.S. Department of the Interior. Report No. PB-250, December 1975 (originally distributed in 1966 by the U.S. Department of the Interior).

Woo, V. 1982. Drought management: expecting the unexpected. *Journal of the American Water Works Association* 74(3): 126–31.

# CHAPTER TWO

# Water-Demand Forecasting

No area of water-resources planning is more beset by risk than the forecasting of water demands. Water facilities are designed, sized, and phased on the basis of projections made by the water forecaster. If future water use exceeds the level forecast, the planned facilities will be extended beyond their economic loading and/or service limits, and water supply deficits may occur. If future water use fails to reach the level forecast, unnecessary economic and environmental costs will have been incurred. In spite of the risks involved, however, attempts at forecasting must be made (even in the face of uncertainty) if planners are to have some role in managing the future rather than just witnessing its arrival.

Since planning, design, and construction of water facilities proceed slowly, and since most facilities are relatively long-lived, water use is customarily forecast over long periods—20, 30, 50, or even 100 years. Given such a long time frame, errors in water-use forecasts can occur for many reasons. Inappropriate or unintended assumptions may be made in determining the parameters of the forecast; for example, future population may be incorrectly projected, changes in the mix of household types may be omitted, or changes in the real level of water prices may be ignored. Other errors may arise in the course of determining the relation between the values of these parameters and the level of water use each implies. For example, conservation efforts may alter the amount of water used in future households, even though all other factors remain the same.

As these examples illustrate, a water-use forecast is a conditional prediction of the level of water use at some future time. Forecasts are conditional because they contain assumptions regarding future levels of water-using activities, future relationships between water use and these activities, future economic conditions, and so on. Any particular forecast provides an estimate of the most likely level of future water use, assuming that all of the underlying assumptions prove correct. Accordingly, forecasting methodology is as much concerned with finding appropriate assumptions as with calculating expected water use once the assumptions are identified.

Demand forecasting has both short-term and long-term applications in water-utility management. In the short term, assuming fixed quantities of water resources and capital investment, demand forecasting serves the following purposes (Saleba 1985):

• Financial planning and management—determining how much money is needed and how and when to use it wisely
• Projecting revenue receipts at current rates to determine if and when a rate increase is needed
• Estimating cost of service and setting rates—figuring how much to charge each customer class
• Risk management—assessing how changes in certain factors might affect the utility and its customers

For the long term—a period during which new sources of supply and new facilities might be called for—demand forecasting serves the following purposes (Saleba 1985):

• Developing a long-term financial strategy that involves the least cost to the utility
• Planning the water system—knowing how much water to deliver and when to deliver it in order to develop the water supply and related system
• Setting objectives for rates and policy—using water rates and policies to guide or reflect the use trends of the system and its customers (obtained by developing a good understanding of the types of demands that will be placed on the system)

The importance of accurate demand forecasting is shown schematically in figure 2-1. Average annual water demands are projected by means of methods and techniques that are discussed in this chapter. Seasonal, daily, and hourly fluctuations in demand are projected on the basis of observed flow records or studies of other areas (see chapter 1). Future facility requirements are then determined from projected peak demands. An operating and capital budget is developed to ensure that additional facilities are constructed as needed to

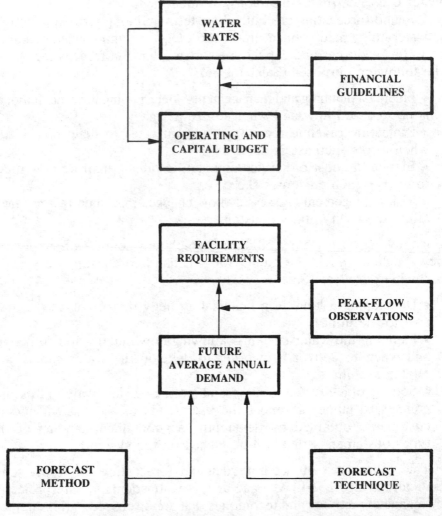

Figure 2-1. Flow chart of water-demand forecasting. (Adapted from figure on p. 1 of course outline for AWWA seminar on forecasting techniques, San Antonio, Texas, May 21–22, 1986; reproduced by permission, copyright © 1986, American Water Works Association)

meet increases in peak water demands. Financial guidelines adopted by the utility are used to set rates for each class of service, ensuring that sufficient revenues are generated to finance needed facility improvements.

## DATA COLLECTION

The single most important step in a water-use forecast consists of determining how much data to collect. This decision influences the choice of forecasting method, and it has a major impact on the quality and credibility of the forecast results. The more accurate and complete the data used are, the more reliable and useful the forecast will be— and the more efficient and productive the plans and projects based on the forecast will prove to be.

### DEFINING THE STUDY AREA

The first step in every water-use forecast is to define carefully the geographic area to which the forecast applies. This step is required even when the forecast is part of a larger water-resources planning activity, because the study area adopted for the overall planning study may not be appropriate to the water-use forecast.

In general, the study area consists of the service area of one or more water utilities. Boland et al. (1983) suggest the following deviations from this definition:

1. A water utility may anticipate expansion of its service area during the forecast period; and in areas with positive economic growth, this is almost always the case. The study area must be expanded, therefore, to coincide with the largest service area expected during the planning period.

2. A portion of an existing water-utility service area may lack public water service at present and may be unlikely to receive it during the planning period. This situation often occurs when service areas are defined as coterminous with political jurisdictions that include large tracts of rural or undeveloped land. It may be helpful, in these cases, to demarcate a study area corresponding to the portion of

the physical service area that is likely to be served by the public system at some time during the planning period.

**3.** The planning context of the forecast activity may require inclusion of self-supplied industrial users of water (not connected to the public supply systems) that are located outside the utility service area. If so, the study area must be expanded to include them. A related instance involves industrial users located outside the utility service area who receive public water-supply service by means of private pipelines. Again, the study area must be defined to include these users.

**4.** Problems may arise in obtaining demographic and socioeconomic data about water-using activities in places where service area boundaries do not coincide with political subdivision boundaries. Two solutions are possible: disaggregation of the demographic and socioeconomic data, to obtain data specific to the service area; and expansion of the study area to correspond to the political jurisdiction for which data are available.

**5.** It is sometimes useful to divide a study area into two or more subareas because substantially more data are available about one subarea than about the other(s). Each subarea is defined according to criteria like those applying to the full study area. Water-use forecasts performed under these conditions are known as *multijurisdiction forecasts*.

## DATA TYPES AND SOURCES

Two general types of data from the study area are needed by the forecaster: records of water-use levels, and records of potential explanatory variables. An explanatory variable serves to explain, in whole or in part, past variations in water use. Water-use forecasts always require data for one or more potential explanatory variables and usually require data for water use itself. The most frequently used explanatory variable is service-area population, although many other demographic, socioeconomic, geographic, climatic, and technological variables can be considered.

Table 2-1 lists factors that are known to affect municipal and industrial water use. These factors identify or suggest many of the potential explanatory variables that may be considered. Since these factors determine water use per individual user, variables that indicate the

**Table 2-1. Factors That Influence Urban Water Demand.**

| Type of Variable | Factors | Type of Variable | Factors |
|---|---|---|---|
| Demographic | Population (households, connections, users, etc.) | Sociopolitical (cont'd.) | Cultural constraints or incentives |
| | Housing density | | Consumer education |
| | Type of housing | | Policy variables |
| | Household size | Climatic | Temperature |
| | Construction grading | | Precipitation |
| | Size of lot | | Moisture deficit |
| | Irrigated area | | Implementation of |
| | Connections to public | | drought-tolerant |
| | sewer | | landscaping |
| | Recreation lake | Technological | Input of raw materials[1] |
| Economic | Income level | | Water-recirculation |
| | Assessed sales value of | | rates[1] |
| | residence | | Inspection and repair |
| | Water-rate structure | | of faulty plumbing |
| | Width and level of | | Leak-detection |
| | price blocks | | program |
| | Employee productivity[1] | | Efficiency of water- |
| | Price of water | | using equipment |
| Sociopolitical | Consumer preferences, | | Distribution pressure |
| | habits, and tastes | | Supply dependability |
| | Legal and political | | Allocation of water of |
| | constraints | | different quality to |
| | | | different users |

[1] Applies to industrial water use only.

SOURCE: Adapted from U.S. Army Corps of Engineers (1983), p. II-2.

number of users (population, number of households, number of connections, number of industrial users, and so on) are also needed.

Water-use models may be estimated from either time-series data or cross-sectional data. Time-series data consist of observations of water use and of explanatory variables made over a number of years at the same location. Cross-sectional data are simultaneous observations of water use and explanatory variables at a number of locations during a single time period. Occasionally, time-series and cross-sectional data are used together, but they are more often treated as alternatives.

Time-series data are, in some respects, superior to cross-sectional data for developing forecasting models. By analyzing time-series data, forecasters can identify trends in water use over time and can develop hypotheses regarding the continuation (or not) of these trends into the future. When using a cross-sectional model for forecasting, forecasters must assume that the functional relationships existing among the variables at the recorded point in time will continue into the future. This tends to limit the applicability of these models to situations in which underlying relationships have not significantly changed over time.

Forecasters have recourse to two major sources of data. First, data may be obtained from state or regional agencies—including water resources boards, employment security agencies, industrial development agencies, departments of community and economic affairs, and planning agencies. Water resources agencies may be able to offer informed assessments of planning for water resources, as well as an inventory of the extent and nature of water resources data within the region. Listings of state and regional agencies can be found in most public libraries. Second, water utilities and local planning agencies within the study area are usually knowledgeable about the availability of local data. Common sources for selected variables are shown in table 2-2.

## DATA-COLLECTION EFFORT

The effort necessary to secure data varies widely from variable to variable and from forecasting application to forecasting application. Some data may be readily available in the offices of the water utility or a state or local agency. In these cases, a telephone call, an exchange of correspondence, or an office visit may suffice to obtain the data. In other cases, data collection may involve such strenuous and time-consuming efforts as manual analysis of water billing records, field surveys of users, and encounters with continual referrals from agency to agency in search of data.

How hard it is to obtain data depends to a great extent on the level of water-use disaggregation chosen. The finer the level of disaggregation (in terms of the size of the service area and the extent of sectoral breakdown), the more difficult the data are to obtain. Often data are only available for the whole state or for a large regional area that bears little relation to the study area. Generally, the smaller the study

**Table 2-2. Data Sources for Commonly Used Variables.**

*Historical Data*

1. Water-use data (municipal and industrial)
   - Water utility
2. Population data
   - State, regional, or local planning agency
   - Water utility
   - Economic development agency
   - City or regional planning agency
   - U.S. census of population
3. Number of households or dwelling units: other demographic variables
   - State, regional, or local planning agency
   - U.S. census of population, housing
4. Number of connections
   - Water utility
5. Climatic data
   - National Weather Service, NOAA, U.S. Dept. of Commerce
   - Dept. of meteorology or climatology, state university
   - Water utility
6. Water and wastewater rate structure
   - Water utility
7. Other economic variables
   - State, regional, or local planning agency
   - U.S. census of population, housing, business, manufactures
   - Real property–assessment agency

8. Policy variables
   - State or local governments
   - Water utility
9. Manufacturing employment, output, processes
   - Local or regional economic development agency
   - State employment agency
   - U.S. census of manufactures
   - U.S. Bureau of Labor Statistics
   - Individual firms

*Projected Data*

10. Population, household size, number of households, etc.
    - State, regional, or local planning agency
    - Economic development agency
    - OBERS projections
11. Economic variables
    - State, regional, or local planning agency
    - OBERS projections
12. Manufacturing employment
    - State, regional, or local planning agency
    - Economic development agency
    - OBERS projections
    - Individual firms

SOURCE: U.S. Army Corps of Engineers (1983), pp. II-4–II-5.

area, the greater the reliance must be on data from local agencies. Nonetheless, state or regional planning agencies should always be contacted, since they may have undertaken one or more regional studies that contain detailed information relevant to the study area.

Disaggregation of data is also a problem when relevant information can only be obtained from federal sources. If federal population projections are the only ones available, for example, and if the study area is smaller than the area covered by the federal projection, the forecaster must rely on personal conjecture to fix the percentage of growth attributable to the study area. An alternative is to extrapolate the growth rate on the basis of historical data for the study area.

## ORGANIZING WATER-CONSUMPTION DATA
## AT THE LOCAL LEVEL

The risk associated with demand forecasting progressively diminishes as more data are collected and analyzed. Increased data analysis leads to a better understanding of water-consumption patterns—a prerequisite for further improvement in the planning and management of urban water resources. This section discusses procedures for developing a database file that can handle water-consumption data efficiently at the local level.

### Database File

Many forecast methods require information on historical water use, disaggregated by use sector (residential, commercial, and so on). Disaggregate forecasts offer the planner more complete information and facilitate consideration of water conservation; yet few water utilities are able to provide disaggregate water-use data. Sometimes the data may be obtained by analyzing billing records; in other cases, there may be no immediate means of classifying customer accounts by user sector. Still, the data can be obtained: with the cooperation of the water utility, meter readers can be called upon to classify accounts, so that billing records can later be analyzed. Several billing cycles, involving separate meter readings, may be required in order to complete this task; but it should be undertaken whenever the prospective improvement in the quality of the forecast would seem to justify it. The water-use forecaster is responsible for determining the data collection effort needed in each planning situation.

Many urban water districts have computerized their water-billing procedures. With a little effort, a computerized billing system can be expanded to include data on consumption patterns and other variables needed by the forecaster. In general, an appropriate database file categorizes water-consumption data into three areas: class of use, place of use, and time of use.

**Class of Use** The definition of *use classes* depends on the intended applications of the aggregated data and on the water-use characteristics of the study area. The example that follows comes from a computerized database file developed by the Honolulu Board of Water Supply (Yamauchi and Huang 1980). Other water-use classifications could be selected to suit the water-use characteristics of a different study area. The customer-service classifications used by the Honolulu Board of Water Supply contain the following classes of water use:

| Code | Customer Class | Code | Customer Class |
|------|----------------|------|----------------|
| 10 | Single-family residential | 62 | U.S. nonmilitary |
| 20 | Multifamily residential | 63 | State uses |
| 30 | Commercial | 65 | City and county uses |
| 40 | Hotel | 70 | Agriculture |
| 50 | Industrial | 80 | Religious institutions |
| 61 | U.S. military | | |

**Place of Use** Disaggregating water-consumption data by place of use allows correlation of water use with variables of both supply and demand. When this type of disaggregation is used in conjunction with census tract data, water use can be correlated with several socio-economic variables.

On the demand side, water-consumption data can be classified according to water service districts, U.S. census blocks and tracts, and subdivisions. Census records can then be examined for information on construction dates of buildings and homes, average number of rooms, mean property value, and average family income. On the supply side, water-consumption data can be classified according to water-pressure system and pressure level. The location codes can even be tied to specific pipe nodes in hydraulic models, allowing integration of network analyses and planning studies.

The Honolulu Board of Water Supply uses an eight-digit coding system to define water-use locations. The first three digits identify the census tract; digit four gives the service area; digit five identifies the pressure system; digits six and seven are a computer compilation of subtotals; and digit eight identifies the pressure level.

**Time of Use**   Precise information on time of use is limited by meter-reading and billing schedules. In general, large-volume users are billed monthly, and small users are billed bimonthly. From these records, average daily, total monthly, and annual consumption figures can be compiled. In computerized billing systems, historical data drawn from records of specific time periods can be compiled, enabling the forecaster to analyze water-consumption patterns with respect to time-dependent variables such as climatic conditions and development phasing.

In many cases, detailed information on instantaneous hourly and daily consumption patterns is not available for disaggregated areas because meters are not read over such short time periods. More detailed information can be obtained through metering of individual services and distribution mains. Also useful are detailed records of reservoir storage levels; these are not usually coded for computer storage and retrieval, but are used for manual processing and analysis.

## Forecast Applications

Developing a comprehensive database file can prove very helpful in making short- and long-term projections of water-supply requirements. Other potential applications include: integrating water-pumping and water-consumption patterns to reduce peak-load demands on the water distribution system; and generating timely information for policy-related matters such as pricing, repayment, and cost distribution.

In pursuing this goal, the forecaster should use a collaborative approach that integrates demand-forecasting variables with mathematical modeling of the distribution system and water-pricing policies. The approach taken by the Honolulu Board of Water Supply focuses on the following steps:

**1.** Design a database file for water-consumption data according to class of use, place of use, and time of use.
**2.** Apply alternative statistical procedures for analyzing historical and current time patterns of hourly, daily, monthly, and annual water-consumption data by water-use classes and locations.
**3.** Analyze the economic incentive effects of alternative demand rates on the time patterns of water consumption and on revenues for different use classes and for different locations.
**4.** Evaluate the relationships between alternative demand-rate structures and between revenues with alternative water-supply ar-

rangements and costs, through development and application of a mathematical model that integrates demand and supply systems.

## FORECAST METHODS

A complete review of water-use forecasting approaches would reveal a great many individual methods and techniques, ranging from simple to complex. Some forecasting approaches are conventionally applied to subjects other than future water use, and they will not be considered here.

Pure judgment forecasts, where predictions of future water use are founded on the subjective judgment of one person, constitute simple prediction: no attempt is made to explain present or future water use, and no formal model is developed. Similarly, collective judgment forecasts rely upon the judgment of a number of individuals who have reached consensus by some means. Purely judgmental approaches will not be discussed further here, but it should be pointed out that all forecasting relies to some extent on subjective judgment. Still, although judgment is an essential element in any forecasting method, its role in the forecast ought to be appropriate and explicit.

Some forecasts attempt to explain water use solely in terms of the passage of time. Simple time extrapolation considers only past water-use records: past changes in water use over time are extrapolated into the future. The extrapolation may be accomplished by graphical or mathematical means; and the change over time may be treated as linear, exponential, or logistic, or as conforming to some other functional relationship. In any case, the change in water use observed in the historical record is assumed to define the change in water use that will occur in the future. There is, however, no empirical reason to expect such a relationship to be stable.

Aside from purely judgmental approaches and simple time-extrapolation techniques, water forecasting methods fall into four general categories:

- Single-coefficient methods
- Multiple-coefficient methods
- Micro time-series methods
- Contingency tree methods

The characteristics of these four categories are summarized in the next few sections.

## SINGLE-COEFFICIENT METHODS

Single-coefficient methods employ a single explanatory variable. The most commonly used variables are population and service connections, although any factor affecting water use can be used.

### Per-Capita Methods

Of all single-coefficient methods, the per-capita approach is by far the most widely used. This approach estimates future water use as the product of the projected service area population and the projected value of a per-capita water-use coefficient. Although population may be projected by various means, it is usually obtained from a more holistic econometric forecast. The per-capita coefficient may be taken as fixed over time or it may be projected to change with time. Its value and (where applicable) its rate of change may be determined from past water-use patterns in the same area, in similar areas, for the whole region, or for the whole nation. Alternatively, the coefficient value may be obtained from reference works or from other studies, or it may simply be assumed.

The per-capita method requires relatively little data, and these are usually readily available. Changes in water use are assumed to be explained by the variable of population alone, with possible provision as well for changes over time in unit use. Focusing exclusively on population in this regard carries the implicit assumption that all other factors affecting water use—such as employment, housing type, family income, water price, and weather—will continue to bear the same relationship to service area population that they did in the past. This is a highly questionable assumption, however, particularly in rapidly changing suburban areas.

### Per-Connection Methods

Another type of single-coefficient method uses number of connections (customers) in place of population as the explanatory variable. The advantage of per-connection methods is that historical data on number of connections to a water supply system are more readily available and more accurate than data on past population, which often must be roughly apportioned to the service area and interpolated between census years. The number of connections correlates closely with the number of household units, which in turn correlates better with water use than total population does.

In addition to offering increased accuracy, per-connection methods enable projections to be disaggregated by type of use or customer class. This is particularly important if estimates of the effectiveness of water-use restrictions are liable to vary by customer class and type of use. Per-capita analysis is incapable of dealing with this problem.

## Other Methods

Other methods based on a single explanatory variable are generally used to forecast specific types of water use. For example, commercial and industrial water use may be forecast as a function of the number of employees or of the amount of retail sales. Residential water use may be forecast as a function of the number of households. In each case, water use is assumed to vary proportionately with changes in the selected explanatory variable, and the coefficient value is assumed to change continuously from the past to the future.

Although single coefficient methods are the most widely used of all methods, they have serious shortcomings in most forecasting applications. By limiting the number of explanatory variables to one (per-capita, per-connection, or some other), users of these methods must omit many other factors known to affect water use (see table 2-1). Where the causal relationship between water use and the chosen variable is strong (as, for example, in the relationship between industrial nonprocess water use and industrial employment), the underlying assumptions may be borne out in the short range, but they become more tenuous in the medium-to-long range.

## MULTIPLE-COEFFICIENT METHODS

Multiple-coefficient methods express future water use as a mathematical function of two or more explanatory variables. The variables are chosen because of their past correlation with water use; any number may be included, although working with more than five or six is unusual. In order to forecast water use, the planner must estimate the values of the explanatory variables statistically—usually by means of a regression analysis.

Methods that incorporate more than one explanatory variable fall into one of two categories: those using requirements models, and those using demand models. Requirements models include variables having significant correlations with water use; these variables do not necessarily include price of water or household or per-capita income. Demand models, on the other hand, are based on economic reasoning and only include variables that are: (1) expected to be causally related to water use; and (2) found to be significantly correlated with water use. Demand models include variables of price and income, as well as other variables. The number and nature of explanatory variables actually used in these models may vary greatly from one application to another, depending on such matters as data availability, required accuracy, and local conditions.

Data requirements for multiple-coefficient models may be considerable, depending on the number of explanatory variables used, and some types of data may be difficult to collect. Of course, data-collection efforts must be balanced against potential improvements in the reliability of the forecast.

## MICRO TIME-SERIES METHODS

Time-series analysis is based on the assumption that future water use can be predicted solely on the basis of past water-use patterns. Hence, these models relate future water use only to past levels of water use; no other variables are considered.

Some forecasters, such as Danielson (1979) use less aggregated data and substitute time-series data for cross-sectional data in estimating individual and household demand relationships (both of which are a function of variables that change over time). This approach, and its variations, collectively termed *micro time-series methods,* use either single or multiple explanatory variables. Ideally, water-use data is collected individually for a number of households over a span of time so that changes in the water-use behavior of households can be analyzed over time and between households.

Selecting a data base to use in a micro time-series projection is critical because the data describing the relationships between water use and the explanatory variables are collected from a relatively small number of households that must be representative of a much larger group of users. Households scheduled for examination should be disaggregated into single-family and multifamily classifications. Further breakdown into density classifications (expressed in terms of dwelling units per acre) may also be desired. Since water consumption may be tracked on a daily basis, the size of the data base must be manageably small. All households selected should be free of any unusual water-use characteristics, such as disproportionately large or small lot size, uncharacteristic landscaping (such as cactus gardens), or extensive remodeling. A field survey and/or mail questionnaire is often needed to establish that the households selected are representative of the study area in question. In the Danielson study, households were dropped from the data base if the family occupying the home moved during the sampling period or if a complete set of data was lacking for some other reason.

Census block data can frequently be used in a data base for micro time-series projections. Census blocks are well-defined, tend to include households of similar size and age, and often are small enough to allow time-series analysis. An added bonus is the availability of additional data compiled by the Census Bureau, such as number of bedrooms, number of bathrooms, median family income, occupancy (renter- or owner-occupied), and median market value of the property.

Once a data base has been selected, the water-consumption data available should be adjusted to account for water-billing cycles. Cities normally read only a portion of all the water meters during a single period, enabling the meter-reading staff to be used efficiently. Once taken, meter readings are compiled, and (typically) all customers are billed on a one-month or two-month cycle. Some types of seasonal water use, such as in irrigation or construction, may result in meters' being read periodically but not according to any set schedule. These seasonal uses of water typically are billed irregularly depending on the intensity of water use and the manpower available to take the readings. Thus, water-consumption data gathered for water-billing purposes may not reflect actual usage over the same calendar dates for all customers.

To account for vagaries in water-billing cycles, data on all time-series variables relating to a given customer should initially be obtained daily to ensure that the observation data on the explanatory variable represents the same calendar period covered in the water-meter readings. These daily readings should be taken for at least two consecutive billing cycles. If not enough manpower is available to record the data daily during this initial period, the data can be manipulated (such as by shifting them backward one billing cycle to correspond more closely to actual water consumption), but all such manipulations tend to erode the reliability of the final projection. Once compiled, the data base should be updated at each billing period to track changes in the explanatory variables over time.

Once the appropriate adjustments have been made to the data base, relationships between water use and the chosen explanatory variables are derived just as they are in the single- and multiple-coefficient methods. Using micro (household) time-series data, however, provides estimates of how specific households respond over time, rather than estimates of how individuals in different areas respond to spatial changes in explanatory variables.

In addition to establishing a representative data base, the forecaster must choose a time period that allows sufficient variation in each variable of interest. Time-series data have a long gestation period. It may take from 5 to 15 years to accumulate sufficient data to support reliable time-series analysis. Time-series data are also fragile. Changes in data-collection procedures can disrupt the consistency of a series before it matures, rendering a large portion of previously collected data useless. Changes in the magnitude of a variable's rate of increase or decrease can have a similar effect. For example, if the real price of water has varied by only about 20 percent during the period for which data are available, the effects of an expected price increase of 100 percent or more may be difficult to forecast reliably. At the other extreme, some explanatory variables, such as residential lot size, may remain unchanged throughout the historical period, providing no hint as to how water use will respond to expected future changes.

DeKay (1985) proposes two tactics to mitigate the effects of these potential problems. One is to collect data that are sufficiently disaggregated to allow to introduce modifications (as necessary) to finesse the effects of the changes. For example, data inconsistencies caused by nonuniform billing cycles can be overcome by aggregating monthly data into consistent peak and off-peak periods for each year exhibiting variations in billing procedures. If a large group of one type of user is transferred from one demand classification to another, the damage can be repaired by treating the two classifications involved as a single combined group. The second tactic is to diversify the available data bases. For example, the forecaster can maintain a few years' worth of annual billing records that identify use by census tract, allowing development of cross-sectional data bases for long-term forecasting models in lieu of a long span of time-series data. Such data can also be used to calibrate forecasts for geographic subareas.

Often, implementing these tactics requires collecting data in a more disaggregated form than perhaps seems necessary for analysis at the time of the original forecast. Besides allowing more detailed analysis and the application of specialized techniques, foresight of this kind protects the integrity of the time-series data. Where data cannot be collected over a long enough period to yield statistically reliable estimates of model coefficients, the forecaster may have to collect cross-sectional data in order to achieve the necessary range of conditions. Meter-reading practices during the study period should be examined so that any data limitations caused by irregular or nonexistent billing cycles will be exposed.

## CONTINGENCY TREE METHODS

Each assumption and explanatory variable projection contained in a water-use forecast represents a condition likely to occur in the future. Sets of assumptions and projections are collectively termed *alternative futures*. In some cases, the likelihood of a number of alternative futures may be considered roughly equal, with no obvious "most likely" choice standing out. Such a situation would exist where a range of policy options is to be investigated. The forecaster can respond to this by preparing a number of alternative water-use forecasts, each corresponding to a unique set of assumptions and projections. By this means, the sensitivity of future water-use computations to various combinations of assumptions can be determined, the level of uncertainty inherent in the forecast will be revealed, and a conclusion about probable future water demand can be drawn.

One of the first applications of the alternative futures approach to water-demand forecasting was proposed by Whitford (1973). Called the *contingency tree method,* it projects alternative futures on the basis of various nonreversible events that might occur in the future, altering the demand for water. Subjective estimates are made of the effect each event would have on water use and of the probability of its occurrence. A baseline forecast is prepared, using one of the previous methods, in which the highest-probability outcome of all the postulated events is assumed. The baseline forecast is then modified to illustrate the effect of every possible outcome of the uncertain factors, one combination at a time; finally, the joint probability of each combination is associated with the water-use forecast calculated for that combination.

In addition to assembling the data used in the baseline forecast, the forecaster must identify the factors likely to affect future water use and must make subjective estimates of their effects on water use and their likelihood of occurring. Some analysts, such as Boland and Malloy (1973), have suggested that these estimates give a misleadingly precise description of the variability of water-demand forecasts. Other analysts, such as Collins and Plummer (1974), point out that the contingency approach does force planners to stop putting all their eggs in one basket by focusing on selecting one forecast. At least the contingency tree method encourages the planner to estimate the degree of flexibility appropriate for large-scale, long-term planning of a particular system.

## SUMMARY EVALUATION OF FORECAST METHODS

Forecasting methods can be selected according to their data requirements. In general, the greater the amount and range of data, the more advanced the forecasting method can be. Data type and availability, in turn, are functions of local conditions and the level of data-collection effort. In principle, virtually any kind of data can be obtained with sufficient effort; but the data-collection effort expended in each application should reflect a balance between the costs of collection and the benefits derived from better and more reliable forecasts.

Forecasting methods can also be selected according to the intended application or final use of the forecast results. If the application requires only a forecast of average annual aggregate water use, almost any method will work, subject to data availability and conformity with study area characteristics. If the forecast is to be used in designing treatment and conveyance works, however, methods capable of producing reliable estimates of maximum-day water use may be required.

Likewise, designing a surface-water reservoir may—depending on the project's size and purpose—require forecasts of seasonal water use and maximum-month use, as well as average annual water use. In other cases, alternative forecasts may be essential so that the effects of various levels of economic development on water use can be shown; the forecast method chosen must include appropriate economic measures as explanatory variables. In addition to making good use of available data and information, a well-chosen forecast method will provide all necessary water-use information to those involved in the planning process.

When water-use forecasts are prepared for purposes of project evaluation, the effect of the project on future patterns and levels of water use must be considered. In general, planners must develop forecasts of future water use under two alternative scenarios: one that assumes the absence of the project (the "no-project" alternative); and one that assumes the presence of the project as specified (the "with-project" alternative). In the case of the no-project alternative, future water needs must be met by existing facilities, locally planned additions or replacements, locally implemented conservation measures, and so on. Often, the water-conservation alternative becomes the basis of comparison, with all or some of the anticipated increase in future demand offset by proposed individual water-conservation measures. The ability to evaluate water conservation as an alternative to increased supply has become an important characteristic of modern forecasting methods.

Table 2-3 presents a summary comparison of the forecasting methods discussed heretofore. In it forecasting methods are characterized according to data requirements, forecast applications, and sensitivity to effects of water conservation. As table 2-3 indicates, data requirements vary considerably, depending on the forecast method used. Single-coefficient methods require the least amount of data—most often readily available aggregate water-production and customer data, as in the case of per-capita and per-connection forecast methods. These data are usually available from water utility reports, census reports, and other public records (see table 2-2). Other unit-use coefficients may require the planner to collect additional data, depending on the explanatory variable used in the forecast.

Multiple-coefficient methods require more data collection; the quantity of data needed and the degree of difficulty involved in collecting them depend on the particular explanatory variables selected. Data require-

**Table 2-3. Comparison of Forecasting Methods.**

| | Single-coefficient Methods | | | Multiple Coefficient Methods | Micro Time-Series Methods | Contingency Tree Methods |
|---|---|---|---|---|---|---|
| Criterion | Per Capita | Per Connection | Unit Use Coefficient | | | |
| *Data Requirements* | | | | | | |
| Quantity needed | Very little | Very little | Moderate | Moderate to large | Moderate to large | Depends on application |
| Collection difficulty | Low | Low | Low to moderate | Moderate to high | High | Depends on application |
| *Applications* | | | | | | |
| Preliminary studies | Yes | Yes | Yes | No | No | No |
| Project planning | No | No | When used in disaggregate forecasts | When used in disaggregate forecasts | With adequate data base and time period | Yes |
| Water conservation effects | No | No | When used in disaggregate forecasts | When used in disaggregate forecasts | With adequate data base and time period | Yes |

Source: Adapted from U.S. Army Corps of Engineers (1983), p. III-1.

ments may include demographic variables (family size, family composition, housing characteristics, and so on), economic variables (water prices, family income, home values, and so on), climatic variables (precipitation, temperature, and so on), and other variables, aggregated or disaggregated by use sector.

Micro time-series methods tend to be data-intensive, since detailed information must be collected from a relatively small number of households. Because such data frequently are available only for large use-sectors, they usually must be obtained through field surveys or mail questionnaires.

Contingency tree methods can be applied to forecasting at any level of data requirements. The scope of potential applications, however, is restricted in situations involving low levels of data availability.

In general, the more comprehensive a forecast is, the greater its applicability to long-term project planning studies becomes. Simple methods—single-coefficient methods, for example—are best suited for use in preliminary studies only.

The per-capita and per-connection methods, with their omission of most explanatory variables and their necessarily aggregate nature, are poorly structured to determine the effectiveness of individual water conservation measures; as a result, using them hinders the full evaluation of water conservation as an alternate source of supply. Other unit-use single-coefficient methods and multiple-coefficient methods, when used in disaggregated forecasts, allow assessment of the effects of water-conservation practices in isolation from other explanatory variables. Contingency tree methods are best-suited to evaluating the impact of water-conservation practices, subject to the forecaster's estimates of these practices' effects on water use and their probabilities of occurrence.

## FORECAST TECHNIQUES

The forecasting methods discussed in the previous section rely on the forecaster's ability to estimate the relationship of one or more independent variables to one dependent variable—water consumption. This section discusses several statistical techniques the forecaster can employ when using these methods.

Four primary forecasting techniques have been described by George (1985) as part of an overview of primary techniques used by electrical and natural gas utilities to predict energy consumption and demand: end-use forecasting, time-series forecasting, econometric forecasting, and hybrid models incorporating end-use and econometric techniques. Similarly, Saleba (1985) has identified four major forecasting techniques currently used by the water industry: end-use forecasting, time-series forecasting, econometric forecasting, and normalization (a useful technique for dealing with random and unpredictable data).

## END-USE FORECASTING

End-use models base the forecast of water demand on a forecast of uses for water. Saleba identifies the following elements as necessary for end-use forecasting models:

- A comprehensive list of major relevant end uses for each type of customer (one entry on such a list would be the number of dishwashers for all residential consumers in the study area)
- A forecast of any new end uses that may arise during the planning period
- Calculation of the useful lives of existing and forecast end uses
- Records of changes in water-use efficiencies for each listed end use, plus other factors as needed

George cites two major advantages of end-use models: they are excellent tools for determining the impact of policies and regulations affecting water use; and they are easy to understand. For example, the impact of new housing regulations that require the use of low-flow devices can be estimated by figuring a per-unit impact of the new devices, predicting the number of program participants (this is usually done outside the model), and establishing a formula for phasing the new devices into the system during the planning period. Electrical utilities, in particular, adopt end-use models for program planning and policy analysis, even when the situation does not explicitly involve long-term forecasting. In addition, the fundamental equations of an end-use model are simple and straightforward. Most people can understand the relationship between low-flow shower heads and reductions in water consumption.

George also cites two major disadvantages of end-use models: they are expensive to develop and maintain; and they do not adequately predict the impact of changes in rates or other economic variables. For example, the requisite data for a large end-use model might fill hundreds of computer pages. Such detailed information is typically gathered through periodic customer surveys, which are expensive and time-consuming to conduct. As a result, constructing and maintaining a large end-use model often exceeds a utility's resources. Moreover, end-use models are generally not useful for short- to medium-term forecasting because patterns in end uses do not change quickly.

## TIME-SERIES FORECASTING

Like trend extrapolation models, pure time-series models forecast the future based solely on historical values of the dependent variable. It is a direct forecasting method, unlike the end-use technique, in that it predicts water-consumption levels directly, without resort to various factors that affect water consumption. Time-series models have been used by electrical utilities to estimate energy sales, peak demands, and load shape.

Many different types of time-series models can be used to forecast water demand, some of them quite complicated. The most common time-series technique used by utilities is the Box-Jenkins or ARIMA (AutoRegressive Integrated Moving Average) model. The following example of a Box-Jenkins model is described by George (1985):*

$$E_t = a_0 + \sum_{i=1}^{p} a_i E_{t-i} + e_t + \sum_{i=1}^{q} b_i e_{t-i} \qquad (2.1)$$

where

$$E_t = \text{energy consumption in period } t$$
$$a_i \text{ and } b_i = \text{model parameters to be estimated}$$
$$e_t = \text{unobservable error terms in period } t$$
$$p \text{ and } q = \text{length of the behavioral effect}$$

* This equation appears in George's paper (on page 17) as equation (1). $E_{t-1}$ and $e_{t-1}$ have been changed to $E_{t-i}$ and $e_{t-i}$ to correct errors in the original.

The $a_i$ parameters are called *autoregressive parameters,* and the $b_i$ parameters are called *moving average parameters.* More complex models can be defined using transfer functions from one or more independent variables to the variable of interest. In addition to containing the lagged values of the explanatory variable, a transfer function essentially includes current and lagged values of selected explanatory variables in the analysis.

George cites the following advantages of time-series models:

**1.** They are the least data-intensive of all models and therefore are relatively easy to use.
**2.** They can be developed by means of various standard software packages, and the skill necessary to develop simple models is moderate.
**3.** The forecast accuracy of these models is good in the short run, which is their most frequent application by utilities.
**4.** They are convenient to use when seasonal or hourly patterns of demand must be predicted.

George cites two major disadvantages associated with these models:

**1.** They are not useful as policy tools, and they are inaccurate when significant changes in determining variables occur in the future.
**2.** They can be quite sensitive to their starting values, which carry the greatest weight in the forecast.

As to the first disadvantage (low accuracy), Saleba states that time-series models are only valid for the short term—with a maximum durational validity of perhaps two years. Therefore, in a utility setting where change may occur very slowly and long-term planning is crucial, time-series models may not be particularly useful by themselves. Even without such changes, forecast accuracy is low in the long term.

As to the second disadvantage (sensitivity to starting values), water-demand forecasts would be most affected in models treating climate as the independent variable. If the most recent historical climatic values represented an unusually mild or cold period, the forecasts would differ significantly from forecasts based on readings from a more typical climatic period.

## ECONOMETRIC FORECASTING

Econometric forecasting, also referred to as *regression analysis,* originated with the work of Francis Galton. The studies of genealogical inheritance inspired by Darwin's work led Galton to believe that everything could be studied quantitatively. One of his studies involved the linear trend between the heights of fathers and their sons; the slope of the trend line on this particular study was positive but less than 1, so Galton called the relationship a "regression toward the mean." The term *regression* has since been applied to any linear trend (Dowdy and Wearden 1983).

Econometric forecasting uses regression analysis and economic theory to determine the structural interrelationships among variables. George classifies models as econometric if they represent the key determinants of demand econometrically, whether at the aggregate level or the end-use level. In contrast, models are classified as end-use if they predict demand at the end-use level and if key behavioral relationships are determined exogenously to them or by noneconometric algorithms; an example of this is life-cycle cost analysis.

Saleba describes the econometric approach as consisting of a statistical estimate of the historical relationships between water consumption and different factors (the independent variables), with the underlying assumption that those relationships will continue unchanged in the future. Econometric forecasting is an indirect technique, in the sense that the demand forecasts are derived from forecasts of the independent variables.

Nonetheless, the distinction between econometric models and end-use models is somewhat artificial. For one thing, econometric modeling is used in estimating data and parameters for end-use forecasting models. And while econometric models are generally used to forecast water consumption for relatively broad aggregate groups of customers (for example, residential, commercial, or industrial groups), instances of econometric models that forecast consumption at the end-use level do exist—especially in energy forecasts (space heating, lighting, refrigeration, and so on). Conversely, models characterized as end-use often have econometric equations imbedded in them.

A typical econometric model consists of an equation or set of equations linking demand to determining variables such as price, income, employment, and weather. A very simple econometric model is represented in the following equation (George 1985):*

$$E_t = a_1 + a_2 P_t + a_3 Y_t + a_4 DD_t \qquad\qquad (2.2)$$

where

$$E_t = \text{water consumption in period } t$$
$$P_t = \text{price of water in period } t$$
$$Y_t = \text{income in period } t$$
$$DD_t = \text{degree-days in period } t$$
$$a_1, a_2, a_3, a_4 = \text{fixed but unknown constants}$$

The model in equation 2-2 is linear, but econometric models have been estimated in a wide variety of linear and nonlinear forms. A popular model specification expresses both the dependent variable and the independent variables in logarithmic terms. In a model like this one, the parameters of the equation represent elasticities that measure the percentage change in the dependent variable, given a 1 percent change in an independent variable such as price. The final choice of functional form for an econometric model is based on such factors as how accurately the form represents the underlying relationships, how easy it is to interpret, and whether it is computationally feasible.

George cites the following advantages econometric models have over alternative models for forecasting water consumption:

**1.** They can explicitly model most of the underlying influences on demand, given the appropriate data.
**2.** They are based on an underlying theory of customer behavior.
**3.** They are less data-intensive than other models and therefore can usually be developed more quickly and inexpensively.

To these advantages, Saleba adds the fact that econometric techniques are valid for both long- and short-term forecasts.

Econometric models assume that the relationships between the dependent variable and the independent variables (reflected in the model parameters) remain the same in the future as they were in the past;

* This equation appears in George's paper (on page 17) as equation (3).

but they do not assume that the dependent variable will grow at the same rate in the future as it did in the past. Econometric models can also ignore the technological characteristics of water-using equipment and the detailed assumptions about customer behavior—all of which are implicitly reflected in the model parameters. The only requirement is that equipment and customer characteristics not change dramatically in the future.

The following potential disadvantages of econometric models are cited by George:

**1.** They sometimes result in aggregated bias because observations are used for groups of customers rather than for individual consuming units.

**2.** They generally cannot predict the impact of exogenous factors affecting demand that are not reflected in the historical data.

**3.** They must deal with the vagaries of real-world data.

An example of a model subject to aggregated bias is a model based on data for two-digit SIC system industries. Econometric models implicitly assume that the data represent homogeneous consumers. In fact, however, two-digit industries can be quite heterogeneous, and models based on observations at this level of aggregation can produce biased parameter estimates and erroneous predictions. Examples of exogenous factors that are difficult for econometric models to reflect can be found in many energy- and water-conservation programs and in government-mandated building and equipment standards. Finally, problems such as heteroskedasticity, autocorrelation, and multicollinearity can result in erroneous estimates of model parameters or, in the case of multicollinearity, can mean that statistically significant estimates for specific variables cannot be made at all.*

In spite of these and other problems, econometric models remain a valuable tool for the forecaster. Econometric modeling is one of the few ways of accurately reflecting the impact of price changes on demand forecasts, and it will continue to be used by utilities either alone or in combination with other modeling techniques.

---

* Heteroskedasticity occurs when the variance in the model ($e_i$, in equation 2.2) is not constant across observations. Autocorrelation arises when the disturbance occurring at one point of observation is correlated with the disturbance at other observation points. Multicollinearity occurs when two explanatory variables are so closely correlated that the separate effects of each cannot be distinguished on the basis of the available data.

## Step-by-Step Procedure

Saleba (1985) developed the procedure that follows for developing an econometric demand forecast.

### Step 1: List the Factors Affecting Water Consumption.

The most fundamental and conceptually important step is to develop a thorough understanding of the system's demand characteristics. Substeps that should be taken by the forecaster include the following:

**a.** *Understand the local economy.* What type of business is prevalent in the study area? Does it use a lot of water? Does it employ a lot of people? How does it affect housing development, local incomes, and so on? What are the economic projections for the study area?

**b.** *Analyze the demographics.* How many people are in the study area? How many are moving in, and how many out? What is the average household size? Are people living in single- or multiple-family housing? Where in the study area is the development or lack of development taking place?

**c.** *Study trends in water consumption.* Are people voluntarily conserving water? Are they buying more water? Are they using more appliances? Are they having small or large lawns/gardens? Are any developing businesses water-intensive?

**d.** *Correlate water consumption with weather patterns.* What types of weather lead to higher or lower water consumption? Is temperature or precipitation (or both) a determinant of peak water consumption? Does weather differ at different places within the study area?

**e.** *Locate the water demand.* What is the geographic distribution of water consumption in the study area? Where are the water-intensive businesses and the single-family residential areas located? How does the geographic distribution of water consumption mesh with the geographic distribution of likely development in the study area?

**f.** *Understand your customers' water-consumption response to rates.* How have rates changed in the last several years? Have the changes affected water consumption?

**g.** *Divide water customers into groups that share the same or similar water-consumption characteristics.* How do residential water-consumption patterns differ from commercial ones? What distinct types of commercial, institutional, or industrial customers have the same water-consumption characteristics?

**Step 2: Estimate Historical Relationships.** A reliable and detailed history of water consumption is critical to the econometric forecast. For each group of customers identified in Step 1(g), water-consumption data should be gathered for as long a historical period as possible and for as short a time interval as possible. For example, a forecast built on daily water-consumption records for twenty-five years or on weekly records for ten years will carry a high degree of credibility, whereas a forecast built on two years of annual records may carry no more weight than a simple extrapolation of historical trends.

Once the historical water-consumption data have been gathered, the next step is to qualify and gather all of the factors identified as having a significant effect on overall water consumption in the study area. For example, if water rates are a significant factor, water-rate data should be collected over the same period and at the same intervals as the water-consumption data are. If new housing starts are a significant factor, then the local chamber of commerce should be contacted for records on housing-start dates.

As an illustration, assume the simple case of a single customer class having identical water-using characteristics (Saleba 1985). Annual meter data for five years are as follows:

| Year | Historical Water Demand (1,000 gallons) |
|------|------------------------------------------|
| 1980 | 1,000,000 |
| 1981 | 1,227,500 |
| 1982 | 1,312,500 |
| 1983 | 1,356,000 |
| 1984 | 1,166,000 |

Price of water and total employment are chosen as the independent variables. Accordingly, the following data are compiled:

| Year | Employment | Price of Water |
|------|------------|----------------|
| 1980 | 100 | 0.125 |
| 1981 | 125 | 0.150 |
| 1982 | 135 | 0.175 |
| 1983 | 140 | 0.185 |
| 1984 | 120 | 0.185 |

**Step 3: Forecast Future Relationships.**  In this step, statistical theory is used to develop a mathematical relationship between water consumption and the independent variables (total employment and price of water). Usually, this can be done on a computer equipped with user-friendly software. In the foregoing example, the estimated equation looks like this:

$$\text{water consumption} = 100{,}000 + (9{,}500 \times \text{total employment}) \quad \text{(2.3)}$$
$$- (400{,}000 \times \text{water price})$$

Equation 2.3 can now be used to forecast future water demand. Once the forecast values of the independent variables are substituted into the equation, the planner simply calculates the equation and solves for the forecast dependent variable (water consumption). An assumption is made here that past historical relationships will hold into the future. This area relies heavily on the judgment of the forecaster and is the area most subject to criticism.

The most difficult part of this step is forecasting the independent variables. Often, the forecasting process requires the planners to make additional assumptions. For example, future water rates are affected by inflation. Therefore, a forecast of water rates implicitly requires a forecast of inflation. The planners may choose from among different forecasts having a variety of sources, using their own judgment about which one is better.

In the example, forecasts of the two independent variables are assumed to be as follows:

| Year | Forecast Total Employment | Forecast Price of Water |
|------|---------------------------|-------------------------|
| 1985 | 120 | 0.195 |
| 1986 | 115 | 0.200 |
| 1987 | 130 | 0.200 |
| 1988 | 145 | 0.215 |
| 1989 | 145 | 0.225 |

These forecast values are used in equation 2.3, and the result is the following series of water-demand forecasts:

| Year | Forecast Water Demand (1,000 gallons) |
|------|----------------------|
| 1985 | 1,162,000 |
| 1986 | 1,112,500 |
| 1987 | 1,255,000 |
| 1988 | 1,391,500 |
| 1989 | 1,387,500 |

**Step 4: Sensitivity Analysis.** From the viewpoint of usability, the fourth and final step—sensitivity analysis—is probably the most critical because it measures how sensitive the model is to errors and changes.

There are two major sources of error in an econometric forecast. The first is the estimated relationship between water consumption and the independent variables. The wrong independent variables may be chosen, or their impacts on water consumption may be incorrectly estimated. For example, suppose that instead of being:

$$\text{water consumption} = 100,000 + (9,500 \times \text{total employment}) \quad (2.3)$$
$$- (400,000 \times \text{water price})$$

the "true" equation for water consumption, measured correctly, is:

$$\text{water consumption} = 300,000 + (3,500 \times \text{total employment})$$
$$- (425,000 \times \text{water price}) \quad (2.4)$$
$$+ (12,000 \times \text{housing starts})$$

Depending on the type of error in the model (equation), the results may be totally inaccurate. Some technical methods exist to judge the strength of the relationship between the independent variables and the dependent variable, and these can be found in a reference book or software program.

The second source of error is the forecast of independent variables used to generate future water consumption. For example, suppose that the original forecasts of total employment and price of water are slightly wrong, and that the true forecasts are:

| Year | "Wrong" Forecast Total Employment | "Right" Forecast Total Employment |
|------|------|------|
| 1985 | 120 | 129 |
| 1986 | 115 | 126 |
| 1987 | 130 | 135 |
| 1988 | 145 | 150 |
| 1989 | 145 | 165 |

| Year | "Wrong" Forecast Price of Water | "Right" Forecast Price of Water |
|------|------|------|
| 1985 | 0.195 | 0.197 |
| 1986 | 0.200 | 0.197 |
| 1987 | 0.200 | 0.210 |
| 1988 | 0.215 | 0.225 |
| 1989 | 0.225 | 0.230 |

As the following data show, even seemingly insignificant errors can result in significant inaccuracies in the forecast:

| Year | "Wrong" Forecast Water Demand | "Right" Forecast Water Demand | % Error |
|------|------|------|------|
| 1985 | 1,162,200 | 1,246,700 | −6.8 |
| 1986 | 1,114,500 | 1,218,200 | −8.5 |
| 1987 | 1,255,000 | 1,298,500 | −3.4 |
| 1988 | 1,391,500 | 1,435,000 | −3.0 |
| 1989 | 1,387,500 | 1,575,500 | −11.9 |

Because forecasts of the independent variables will almost certainly be inaccurate to some extent, and because a utility must be able to set out some plan even in the face of tremendous uncertainty, a forecast should lay out the reasonable likelihood that it will be wrong. There are several ways to do this. First, if forecasts for the independent variables came from different sources, such as the state employment division and an independent consultant for employment figures, separate water-consumption forecasts can be made using both to see how it affects future water demand.

For example, suppose that two different employment forecasts have been developed—one by the state economic development agency, and one by a private firm called ABC Econometrics. The separate employment forecasts can be converted into two distinct water-consumption forecasts using each of these sources. By this means, the forecast price of water remains constant, and the sensitivity of the forecast to total employment can be measured. In the continuing example, the results are as follows:

| Year | State Total Employment Forecast | Forecast Water Consumption | ABC Total Employment Forecast | Forecast Water Consumption |
|------|------|------|------|------|
| 1985 | 125 | 1,209,500 | 115 | 1,114,500 |
| 1986 | 140 | 1,350,000 | 118 | 1,141,000 |
| 1987 | 150 | 1,445,000 | 130 | 1,255,000 |
| 1988 | 175 | 1,676,500 | 150 | 1,439,000 |
| 1989 | 190 | 1,815,000 | 165 | 1,577,500 |

A second method for measuring sensitivity is to develop a set of scenarios—likely sets of interrelated events—and plug them into the forecast equation. For example, suppose that the forecaster wanted to measure the impacts of rapid and sustained economic growth on future water consumption. The first step would be to estimate a new set of independent variables; next, the variables would be plugged into the forecast equation. In this scenario, rapid economic growth would produce a rapid increase in employment and a concomitant reduction in water rates (because there would be more customers to pay the fixed costs). The resulting forecast would be as follows:

### HIGH-GROWTH SCENARIO

| Year | Forecast Total Employment | Forecast Water Price | Forecast Water Cons. |
|------|------|------|------|
| 1985 | 130 | 0.180 | 1,263,000 |
| 1986 | 175 | 0.175 | 1,692,500 |
| 1987 | 225 | 0.173 | 2,168,300 |
| 1988 | 250 | 0.170 | 2,407,000 |
| 1989 | 260 | 0.168 | 2,502,800 |

A third method for measuring sensitivity is to perform what is commonly called a *Monte Carlo simulation* because of its randomness. Under this method, some random error (within specified limits) is attached to some or all of the independent variables, and a water-consumption forecast is generated for each case. The variance in each forecast is analyzed to determine whether or not it converges toward some average over a very large number of tests. Convergence toward an average indicates that the model is a good representation of the historical relationship between the forecasted values and the model parameters.

Because this method requires many runs (sometimes thousands of them) before it makes sense, it cannot be demonstrated here. A computer must be used for Monte Carlo simulation. The method is most appropriate when the forecast includes a relatively random independent variable, such as weather.

## HYBRID MODELS

George defines a *hybrid model* as one that predicts demand at the end-use level but uses econometrics to address at least one of the key relationships in the model. Hybrid models are most often used by energy utilities attempting to make best use of available modeling techniques and data. They generally have more sophisticated choice algorithms than pure end-use models do. At the same time, they are generally better at predicting the impact of government- and utility-sponsored demand-side management programs than pure econometric models are. Hybrid models generally offer a useful way of combining the best techniques available within the budget and time constraints associated with model development.

The disadvantages of hybrid models cited by George are more properly ascribed to individual components of a particular model than to the class as a whole. Because hybrid models often begin as end-use models that are modified over time as better data become available, they sometimes suffer from a lack of efficiency and elegance in comparison to models that are developed according to a present plan. In addition, the ad hoc way in which econometric elements are sometimes incorporated into hybrid models sometimes creates problems of interpretation.

## NORMALIZATION

Saleba describes normalization as a generalizing adjustment of historical data that involves the same steps as forecasting. Historical data are gathered for both the dependent variable and the independent variables, and the relationships between variables are statistically estimated in the form of an equation; then the *normal* or long-term statistically adjusted average values of the independent variables are plugged into the forecast equation to generate the forecast that would result under normalized conditions.

This procedure is particularly useful for models using weather as the independent variable, because weather is so random and unpredictable. When generating a weather-normalized water-consumption forecast, the forecaster must possess historical data that cover a long enough period measured at short enough intervals to allow adequate representation of past weather patterns. Assuming that this is the case, the forecaster simply takes the normal (adjusted) values for the weather variables, plugs them into the forecast equation, and calculates water consumption as if the weather had been precisely normal. In order to do this, of course, the forecaster needs a definition of *normal weather*. Fortunately, the National Weather Service has already established such definitions for many areas of the country by taking a long-term statistically adjusted average of many weather variables, including temperature, precipitation, and cloud cover.

This procedure is also useful for handling any other factor that the forecaster perceives as having something "abnormal" built into it. For example, if a water-consumption forecast were needed for a large industrial user that had been closed by a strike for three months of the preceding year, the independent variables (employment, production, and so on) could be normalized to determine the probable future water consumption that would occur, absent labor trouble.

## SUMMARY EVALUATION OF FORECASTING TECHNIQUES

Table 2-4 summarizes the advantages and disadvantages of the four primary modeling techniques used by utilities. These advantages and disadvantages can vary from application to application, and it is not uncommon for utilities to use more than one approach when developing a forecast—each acting as a cross-check on the other. In a sense, this practice merely acknowledges what all forecasters know to be true: forecasting is inherently uncertain.

George suggests that planners developing a forecasting technique give special consideration to the issues of cost, understandability, and accuracy. Developing and maintaining complex models can require

**Table 2-4. Advantages and Disadvantages of Alternative Forecasting Techniques.**

**Time-Series**
*Advantages*
- Minimal data requirements
- Low cost
- Forecast accuracy generally good in short run
- Can predict seasonal and daily patterns

*Disadvantages*
- Does not treat underlying factors explicitly
- Not useful for policy analysis
- Accuracy low in the long run

**Econometric**
*Advantages*
- Explicitly models underlying influences on energy demand
- Based on explicit theory of consumer behavior
- Less data-intensive than end-use models

*Disadvantages*
- High skill level required to develop models
- Difficult to address some policy issues
- Sometimes difficult or impossible to identify individual variable impacts (e.g., multicollinearity)

**End-Use**
*Advantages*
- Good policy-analysis capabilities
- Relatively understandable

*Disadvantages*
- Often lacks endogenous behavioral component
- Data-intensive
- Costly

**Hybrid**
*Advantages*
- Better behavioral component than pure end-use models
- Better policy analysis capabilities than most econometric models

*Disadvantages*
- Data-intensive
- Costly
- Ad hoc nature can make interpretation difficult
- Can lack efficiency and elegance

SOURCE: George (1985), pp. 28–29; reproduced by permission of ASCE.

the investment of millions of dollars and the continued efforts of several full-time staff members. While such high costs are often justified in light of the magnitude of the decisions addressed by the model, these costs can easily exceed the budgetary limits of many utilities. Utilizing a series of smaller, less-expensive models, each designed for a specific purpose, can sometimes be more cost-effective.

The logic underlying some modeling techniques is easier to comprehend than that underlying others. End-use models, for example— while often complicated and time-consuming to implement—tend to be relatively simple to understand. Econometrics and time-series models, on the other hand, have a black-box reputation and can be difficult to explain to upper management, regulators, and other key decision-makers.

A high degree of accuracy is extremely difficult to obtain, especially in long-term forecasting. Nevertheless, some forecasting techniques have proved more accurate than others for certain applications. The forecaster must strive to use the most accurate means possible, given available resources and practical time constraints.

Saleba notes a common misconception that, because forecasting is imperfect, no forecast or a simple ruler-and-graph-paper forecast is better than a forecast that is sophisticated but wrong. This attitude arose in response to a practice that some forecasters have a hard time giving up: presenting a single point forecast. Forecasts can be wrong, and they usually are. Therefore, it is important to acknowledge where the sources of uncertainty are in a given forecast. To make informed decisions in the face of uncertainty, the decision-maker should know what and how much is at risk. Presenting a range of forecasts is the only way to counter the argument that no forecast is better than a wrong one.

## IMPACTS OF WATER CONSERVATION

A key criterion in selecting a forecasting method is the ability to evaluate water conservation as an alternative to supply. In many cases, the water-conservation alternative can serve as the basis of comparison, with all or part of the anticipated increase in future demand offset by individual water-conservation measures.

This section discusses the role of conservation in water-supply planning. Following a definition of the term *water conservation,* various conservation measures are described for the residential, industrial, and commercial sectors. The section concludes with a discussion of ways conservation can be applied in the forecasting process. The beneficial and adverse impacts of conservation on urban water use are described, and suggestions are offered about how to deal with the uncertainty involved in estimating the effectiveness of water-conservation measures.

## DEFINITION OF WATER CONSERVATION

The definition of *water conservation* has changed and expanded over time, in response to economic conditions, natural disasters, and public attitudes. Since the beginning of the twentieth century, the term has been equated with the practice of storing water in the spring for use during the dry summer months. As a result of droughts in the Dust Bowl during the 1930s and in the Northeast during the 1960s, reliability of supplies has become a key criterion in planning for storage facilities. Public conservation measures have traditionally focused on short-term responses to drought.

Public attitudes began to change in the late 1970s, when widespread water shortages led to increased recognition of the physical and economic limits to development of further supplies. Greater attention was given to measures that modified water use, improved the management of existing supplies, and promised (in conjunction with augmented supplies) to provide for long-term needs.

The modern definition of *water conservation* is any beneficial reduction in water use or water losses; clearly, this definition focuses on the efficient management of water use and existing supplies. Water management practices contribute to conservation only when they meet the following two tests: they reduce water use or losses; and their total benefits exceed their total costs (Crews and Mugler 1982).

## WATER-CONSERVATION MEASURES

Water-conservation measures may be classified as either regulatory practices or management practices (Orange County Municipal Water District 1985). Regulatory practices include all measures taken in response to local, state, or federal legislation. A legislative requirement prohibiting the hosing down of driveways, for example, falls under this category. In general, all systemic responses to regulations or restrictions that carry penalties or sanctions for noncompliance are deemed regulatory practices.

Management practices are voluntary actions taken by responsible agencies or water purveyors that result in beneficial reductions in water use or water losses—either directly or by means of incentives that these actions create for water users. This category of water-conservation measures includes retrofitting households with water-saving devices, leak-detection and repair programs, conservation-motivated metering and pricing programs, tax incentives and subsidies, and educational and promotional programs to stimulate and acknowledge voluntary conservation efforts by the general public.

Potential water-conservation measures can be distinguished by duration of implementation. In most cases, conservation measures are considered long term or permanent: once implemented, they typically remain in effect throughout the study period. Certain circumstances of water supply, however, dictate promulgation of short-term or contingent conservation measures on an ad hoc basis. For example, such measures might be incorporated into a contingency plan for drought management, where temporary reductions in water use would be called for during periods of possible water-supply failure. Since most regulatory conservation measures fall into the short-term category, they are treated as drought-management practices in this book.

This section summarizes the impacts of water conservation in the residential, commercial, and industrial sectors. Since the emphasis is on urban water use, agricultural water conservation is not covered. The use of metering and pricing programs to promote conservation is discussed in detail in chapter 4. Water reclamation, which can be viewed as a water-management practice, is discussed separately in the next section.

## Residential Water Conservation
The conservation measures discussed in this section comprise low-flow water fixtures and appliances, reductions in water pressure, landscape irrigation practices, social attitudes, and leak-detection programs.

### Low-flow Water Fixtures and Appliances
The wide variation in reported savings for low-flow water fixtures and appliances has resulted

**Table 2-5. Expected Water Savings from Fixture-use Survey.**

| Fixture | Basis[1] | Predicted Savings (gpcd) | |
| --- | --- | --- | --- |
| | | Single-family | Multifamily |
| Low-flush toilet | 4.0 flushes/capita/day, low-flush toilet saves 2.0 gal/flush | 7.4[2] | 8.0 |
| Low-flow shower | 4.8 minutes shower time/capita/day, low-flow showerhead saves 1.5 gpm | 4.5[3] | 7.2 |
| Faucet aerator | Restricts flow to 2.75 gpm | 0.5 | 0.5 |
| Dishwasher | New models use an average of 3.3 gal/load less with 0.17 loads/capita/day | 0.6 | 0.6 |

[1] Results from *Survey of Water Fixture Use* report, except for faucet aerator, which is based on other published data.
[2] Takes into account that 20 percent of single-family homes had already been retrofitted to reduce their toilet-water use by an average of 3 gpcd.
[3] Takes into account that 46 percent of single-family homes had already been retrofitted to reduce their shower-water use by an average of 5.8 gpcd.

SOURCE: HUD (1984), pp. 5-7.

in a general lack of confidence in their efficacy. Although testing has established per-unit levels of water use for residential plumbing fixtures and water-conservation devices under laboratory conditions, estimates of actual water savings have been based on varying assumptions about typical duration of fixture use, flow rate, temperature, and frequency of use. Earlier studies demonstrating savings with low-flow devices have generally involved a small number of homes—too few to prove that the measured water savings were statistically significant. Moreover, the homes were fitted with a number of devices, so the savings measured could not be broken down and attributed to individual devices.

In the early 1980s, in an attempt to overcome these deficiencies, the U.S. Department of Housing and Urban Development (HUD) embarked on a nationwide survey of water-conservation measures and practices. Over 200 households participated in the survey of water-fixture use. The results of this survey, which are summarized in table 2-5, gave forecasters a better basis for calculating the effects of water-conservation measures than they had ever had before (HUD 1984).

An important element of the HUD study was the distinction it made between single-family and multifamily conservation savings. The combined impact of water-saving fixtures and appliances amounted to 13.0 gallons per capita per day (gpcd) for single-family homes and 16.3 gpcd for multifamily units. Low-flush toilets (3.5 gallons per flush) and low-flow showerheads (3 gallons per minute (gpm)) were found to offer maximum savings of 8.0 gpcd and 7.2 gpcd, respectively. These savings were applied to multifamily buildings by using the case of old apartment buildings that lack water-conserving fixtures being replaced by new apartment buildings that are equipped with water-conserving fixtures. The older apartment buildings were assumed not to have been retrofitted. In single-family homes, the survey found that 20 percent of the toilets and 46 percent of the showerheads had been retrofitted by the owners. Homeowners who had retrofitted their toilets were saving 3 gpcd, and those who had retrofitted their showerheads were saving 5.8 gpcd. Overall, this partial retrofitting of single-family homes decreases the expectable savings from 8.0 gpcd to 7.4 gpcd for low-flush toilets and from 7.2 gpcd to 4.5 gpcd for low-flow showers.

The savings from installing a faucet aerator (2.75 gpm) were not studied and were assumed to equal those reported in the American Water Works Association's handbook on Water Conservation Management—namely, 0.5 gpcd. The savings from installing a new, water-efficient dishwasher averaged 3.3 gallons per load, which translates into 0.6 gpcd.

One facet of the HUD study of special interest to the water forecaster is its evaluation of advanced water-saving plumbing fixtures. A specially designed dormitory, housing 240 resident engineering students on the campus of the Stevens Institute of Technology in Hoboken, New Jersey, was used as a living laboratory for this purpose. The residence hall has five stories and contains 120 rooms with private baths that include a toilet, tub/shower, and lavatory (similar to those in high-rise residences). The dormitory is equipped with two groups of back-to-back bathrooms: one with conventional water-saving fixtures, and one with advanced water-saving fixtures.

Two products constructed on the basis of advanced technologies were tested. The first was an air-assisted toilet designed to use 2 quarts of water per flush. The second was a showerhead designed to operate effectively at flow rates as low as 0.5 gallons per minute (gpm). The HUD study's results, which are summarized in table 2-6, indicate that these advanced water-saving fixtures can dramatically reduce

**Table 2-6. Summary of HUD Study Results.**

| Water-Conservation Practice | Observed Water Savings[1] | Water-Conservation Practice | Observed Water Savings[1] |
|---|---|---|---|
| 3-gpm showerhead | 7.2 gpcd | Water-efficient | 1.0 gpcd |
| 0.5-gpm showerhead | 13.8 gpcd | dishwasher | |
| 3.5-gallon/flush toilet | 8.0 gpcd | Water-efficient | 1.7 gpcd |
| 0.5-gallon/flush toilet | 19.6 gpcd | clothes washer | |

[1] Compared to conventional nonconserving fixtures and appliances.

Source: HUD (1984), p. 1-7.

interior water use. Compared to water use in a nonconserving bathroom, a savings of 33.4 gpcd is possible; and compared to water use in a conventional conserving bathroom, a savings of 17.3 gpcd can be achieved.

Advanced water-saving fixtures are considerably more expensive and complex than conventional water-conserving equipment. Whether the additional cost and complexity are justified depends on local circumstances. Moreover, consumer acceptance of the advanced fixtures was not universal in the HUD experiment; some students purchased and used hand-held showerheads, which they connected to the faucet of the tub. Presumably, this would not be a problem in an owner-occupied home, where the owner was committed to reducing water use, but it might be a problem in a renter-occupied home or an apartment building. One way to avoid the problem would be to stop installing combination tub/showers, since a separate shower compartment would preclude the use of bypass equipment. No effort was made in the HUD project to address these issues. As a result, the reported water savings may be somewhat overstated.

**Reductions in Water Pressure**  Maximum water flow from a given fixture operating at a fixed setting decreases as water-supply pressure is reduced, because water-flow rate varies inversely with the square root of the pressure drop. For example, reducing the water-supply pressure from 100 pounds per square inch (psi) to 50 psi at a given water outlet causes the water-flow rate to decrease by about one-third. On the other hand, many water-using appliances—such as washing machines and toilets—control the total volume of water used, and the HUD study showed that most people throttle their shower back from maximum capacity to a comfortable range; therefore, reducing water pressure would have little effect on the major water-using appliances in the home. It would, however, reduce water-system leakage, household plumbing leakage, and outside water use to some degree.

Water use for outside irrigation can be expected to drop where pressure reduction is imposed on existing landscaped areas, since the irrigation systems are usually designed to operate on the higher pressure previously in force. Although water use probably would drop as a result, the effect may be highly undesirable. Where a given new area is designed to receive a lower-pressure supply, irrigation system designs can be expected to compensate for lower pressure in such a way that no reduction in total water use is realized.

Effects of changes in water pressure on residential water consumption were evaluated as part of the HUD study. The survey's results indicate that water savings of 6 percent can be expected for new single-family homes designed to operate at water pressures 30 to 40 psi below normal.

Reducing water pressure in existing developed areas is probably impractical. Fire-flow requirements may not be subsidized, and complaints from customers may arise. On the other hand, limiting water pressure in new developments to 50 to 60 psi (where practical) does have water-saving advantages. Reducing water pressure saves money by requiring less power to be used in pumping the water from low-pressure areas. In plumbing systems, high water pressure may increase the likelihood and severity of leaking water heaters, water hammer, dripping faucets, dishwasher and clothes-washer noise and breakdown, and leaking water pipes.

The HUD study demonstrates that, taken together, individual fixture savings can significantly reduce interior water use. Figure 2-2 shows the breakdown (by percentage) of how water is used inside two types of homes: one fitted with nonconserving fixtures, and the other fitted with water-conserving fixtures. Notice the reduced water-use percentages for water-saving showers, toilets, and dishwashers—the fixtures that would be found in a water-conserving home.

**Horticultural Practices** The outdoor component of residential water demand varies from one community to another, depending on climate and community characteristics. Typically, outdoor water use accounts

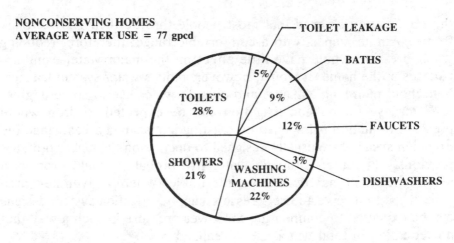

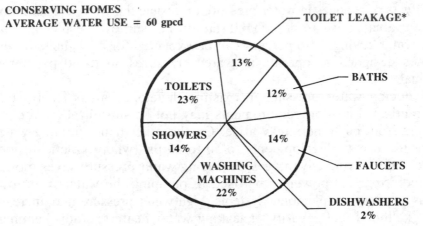

Figure 2-2. Comparison of per-capita interior water use for nonconserving and conserving homes. (Reprinted from HUD 1984, p. 1-3)

for about 50 percent of total residential demand. In addition, outdoor use (predominately lawn sprinkling) is the largest and most important component of seasonal use in the residential sector.

Recent technological innovations in low-cost microcomputers have led to development of a new generation of computerized irrigation-control systems. Unlike existing electrical/mechanical control systems, a computerized system allows for on-site irrigation control from a single central computer that can directly operate all distribution valves.

The higher level of control enables the system to reduce the instantaneous peak demands on the off-site distribution system; and such a reduction in peak flow conserves water, lowers energy consumption, and can avoid costs of facility expansion.

If the computer system is integrated with soil-moisture sensors or other devices, better water management could be attained through the obviation of the psychological tendency toward overwatering. The Los Angeles Department of Water and Power has begun installing moisture-sensing tensiometers at the homes of twenty residents selected to participate in a two-year pilot study. These devices are designed to override the existing automatic sprinkler systems of the participants when the lawn's soil moisture exceeds a certain level. The program is expected to reduce the water needed for residential landscaping by as much as 40 percent (*Journal AWWA* 1985).

Computer control seems likely to reduce annual water consumption for the following reasons:

**1.** It will eliminate water waste, by tailoring irrigation schedules to climatic conditions (putting an end to automatic watering during rainy periods).

**2.** It will ensure that irrigation occurs under optimal conditions, by developing irrigation schedules that are based on measured or computed water requirements of the vegetation.

**3.** It will refine the distribution of water during an irrigating session, by developing separate schedules for drought-tolerant species and slope/turf areas, as well as subpallets where these can be determined.

Decreases in annual water consumption of 30 percent and more have been reported at large-scale landscape-irrigation systems in Southern California (*Landscape West and Irrigation News* 1980). Water-conservation savings of about 20 percent were reported at the Encino Golf Course in the San Fernando Valley after its irrigation system was switched to computer control (*Landscape West and Irrigation News* 1982).

Many studies are currently in progress to identify water-conserving plants for use in semiarid areas, particularly the Pacific Southwest. Called *xeriscapes*—from the Greek word *xeros,* meaning *dry*—collections of these plants form landscapes suitable to an arid region's natural environment. The greatest potential water savings can be realized by modernizing irrigation systems in large water-using areas such as public rights-of-way, park areas, and greenbelt areas in new land developments. In these areas, computerized irrigation control could substantially reduce annual and peak water demands. As more information becomes available, the potential savings achievable by planting drought-tolerant vegetation can be assessed.

**Social Alternatives**   In recent years, water wholesalers have focused their attention on changing the water-use habits of their customers by embarking on extensive public-education campaigns designed to reduce water consumption. These activities were evident in California during the 1976–77 drought, and more recently, in New York. A typical public-education campaign includes the following elements: distribution of a handbook of water-savings ideas; workshops on water conservation; speaker programs; product data on water-saving devices; television and radio advertisements; bumper stickers; plumbing code changes; and distribution of toilet leak-detector kits, plastic bottles for reduced toilet flushes, and showerhead water-flow reducers.

**Leak Detection**   Unaccounted-for water use typically accounts for 5 to 10 percent of total water billings. Depending on the age of the water system and the level of maintenance practiced, pipeline leakage could account for a significant portion of unaccounted-for water use. Remedial steps include metering all lines, checking the operation and accuracy of master and retail meters, recording or estimating hydrant and other municipal uses, and conducting inspections where large leaks are suspected.

### Industrial Water Conservation

The greatest potential for industrial water conservation lies in increasing the use of recycled wastewater in place of freshwater intake. A number of factors may encourage this to happen. Industrial expansion in recent years may overtax available water supplies in areas where growth occurs. Allocations and reallocations of water for other expanding uses, such as irrigated agriculture and thermoelectric power generation, may impose artificial limits on water availability. The cost of fresh water may increase significantly as a result of the additional capital, operation, and maintenance expenditures required to meet stringent effluent-limitation guidelines imposed by federal and state water-quality legislation.

Recycling will have its greatest impacts on cooling water and (to a lesser extent) process water. The impact on boiler-feed water and other uses will be minor, and sanitary water use will hardly be affected at all. In consequence, cooling-water intake will drop sharply, process-water intake will drop moderately, and intake for other uses may actually increase.

## Commercial Water Conservation

Table 2-7 shows an estimated breakdown of water use for fourteen different commercial establishments, as reported by Kim and McCuen (1980). The potential impact of demand modification in the fourteen selected categories is shown in table 2-8. These values suggest that demand modification could reduce water demand in the commercial sector by about 30 percent.

## FORECAST APPLICATIONS

If water conservation is to be incorporated into the forecasting process, the effectiveness of water-conservation measures must be assessed. The effectiveness of a water-conservation measure can be expressed numerically as the quantity of water per unit of time that is saved through implementation of the measure. The effectiveness reading must be obtained from the literature or from engineering analysis of the measure under consideration.

Two difficulties are likely to arise in attempts to transfer estimates from written sources. First, many reported data are not measures of actual results, but are a priori estimates made by other investigators. Even where measures have been implemented and overall reductions in water use achieved, the individual effectiveness of each of the measures may not have been determined. Second, effective data may be reported, not with respect to the affected sector of water use, but as a fraction of some larger aggregate. For example, the effectiveness of lawn-sprinkling restrictions may be given in relation to overall municipal water use, rather than in relation to seasonal residential use. The former result is likely to be inapplicable to a different community, where the structure of municipal water use may be quite different. Unless actual measurements of fractional reductions in water use for the affected sector are available, engineering estimates—either prepared for the purpose or obtained from the literature—must be

Table 2-7. Breakdown of Water Use in Commercial Establishments.

| Commercial Establishment | Toilet Use | Dish-washing | Laundry | Shower | Washing | Misc. | Total |
|---|---|---|---|---|---|---|---|
| Restaurant (gal/chair) | 4 | 16 | | | | 4 | 24 |
| Hospital (gal/bed) | 45 | 30 | 25 | 30 | | 70 | 200 |
| Motel (gal/unit) | 20 | | 5 | 20 | | 5 | 50 |
| Hotel (gal/unit) | 23 | 10 | 5 | 20 | | 10 | 68 |
| Laundromat (gal/machine) | | | 400 | | | | 400 |
| Barbershop (gal/chair) | 10 | | | | 40 | 5 | 55 |
| Beauty salon (gal/station) | 10 | | | | 80 | 5 | 95 |
| School (gal/student) | 4 | | | 1 | | 1 | 6 |
| Office (gal/sq. ft.) | 0.07 | | | | | 0.02 | 0.09 |
| Bank and retail shops (gal/sq. ft.) | 0.04 | | | | | 0.01 | 0.05 |
| Department store (gal/sq. ft.)[1] | 0.04 | 0.03 | | | | 0.01 | 0.08 |
| Department store (gal/sq. ft.)[2] | 0.032 | | | | | 0.008 | 0.04 |
| Service station (gal/sq. ft.) | 0.06 | | | | 0.11 | 0.01 | 0.18 |
| Car wash (gal/sq. ft.) | 0.03 | | | | 4.85 | 0.02 | 4.9 |

[1] With lunch counter.
[2] Without lunch counter.

SOURCE: Reprinted from *Water Conservation Strategies*, by permission; copyright © 1980, American Water Works Association.

**Table 2-8. Potential Water-Use Reductions in Commercial Establishments.**

| Commercial Establishment | Possible Savings of Present Water Use (%) | Commercial Establishment | Possible Savings of Present Water Use (%) |
|---|---|---|---|
| Restaurant | 26.9 | School | 34.5 |
| Hospital | 24.7 | Office | 31.2 |
| Motel | 39.0 | Bank or retail shop | 32.0 |
| Hotel | 34.7 | Department store[2] | 31.2 |
| Laundromat | 30.0 | Department store[3] | 32.0 |
| Barbershop[1] | 21.8 | Service station[1] | 25.4 |
| Beauty salon[1] | 21.2 | Car wash[1] | 20.0 |

[1] 20 percent reduction in washings.
[2] With lunch counter.
[3] Without lunch counter.

SOURCE: Reprinted from *Water Conservation Strategies*, by permission; copyright © 1980, American Water Works Association.

relied upon. Attention should be given to the consequences of error in these estimates; alternative calculations to establish upper and lower bounds of probability are recommended.

Most estimates of water-conservation efficiency are calculated by estimating the difference between water use before implementation of the conservation program and water use after implementation. Hence, the effect of other factors on water use remains unknown. To establish the effect of each measure definitively, a with–without framework is required: what is the water-use pattern *with* a specific conservation measure, and what is the water-use pattern *without* the conservation measure? Only in this way can the true effect of conservation on water use be determined.

Providing predictions of unrestricted water use in each sector affected by a potential water-conservation measure requires preparation of a substantially disaggregated forecast of future water use. Depending on the conservation measure under consideration, separate water-use forecasts may be required for residential use (possibly further dis-aggregated to seasonal and nonseasonal uses), commercial use, industrial use, public and unaccounted-for use, and any other uses that may be the focus of conservation efforts. For best results, these individual forecasts should be prepared as part of a single, integrated forecasting process, so that consistent assumptions are employed throughout.

Using rules of thumb (for example, estimating commercial water use as 15 percent of municipal water use) is likely to cause substantial error, since individual communities vary widely in the proportions of water use attributable to various use categories. Care should also be exercised in the classification of apartment water use; many communities consider such use to be commercial in nature, even though it is responsive to residentially oriented conservation measures.

Water-conservation measures may produce beneficial and adverse impacts that can be identified and described without resorting to full analysis of the water-supply plan. Some of these impacts can be described in monetary terms, while others may appear in forms for which no monetary equivalents are available. In either case, all beneficial and adverse impacts should be identified and should be expressed in monetary (where possible) or other quantitative terms.

## Beneficial Impacts

A reduction in water use implies that certain costs associated with water supply can be avoided. In the case of community water supply, some costs of pumping and treating water vary directly with the quantity of water supplied, so reductions in water use can be expected to bring about reduced costs. Where future water demands will entail augmentation of existing supply facilities, additional monetary savings can be achieved through conservation. These arise because water conservation measures reduce the quantity of water required at any given time in the future, permitting planned facilities to be constructed later than originally anticipated. When construction costs and related costs are deferred, the present-value cost of the capacity-expansion program is reduced. The amount of the reduction is entirely attributable to the water-conservation measure implemented.

On the demand side, an example of beneficial effects that can be estimated in monetary terms is the effect of conservation on energy consumption. A study for the state of Nevada found that significant amounts of energy were saved by water conservation (J. B. Gilbert & Associates 1979). The monetary value of the energy savings often far exceeded the monetary value of the water saved. These savings would be realized by water and wastewater agencies through reductions in water supply and wastewater treatment costs and by homeowners through a reduction in hot-water heating requirements. Conservation measures applied to urban water use also have the effect of reducing the system's peak load, by reducing peak/average ratios.

Water users may also experience nonmonetary benefits, in the form of a sense of satisfaction at doing their part to conserve a valuable and limited resource. While beneficial results of this type are plausible, they have not been generally reported in the literature.

## Adverse Effects

Water conservation may produce a variety of adverse effects on water users—some monetary and others essentially nonmonetary. Where water conservation by industry requires increased attention to process control, the increased effort can be reflected in increased costs. Restrictions on water use for residential lawn irrigation may necessitate relandscaping with more drought-resistant species, with accompanying monetary expense.

Difficulty in estimating implementation costs is most likely to be encountered with measures directed at residential and industrial water use. Various conservation techniques and approaches are possible for these classes of water use, many of them depending to some extent on public-education efforts. While specific measures may entail specific costs, such as the cost of modifying or replacing plumbing fixtures, a multimeasure program may include joint costs, such as those for initiating and promoting the program. In a complex program, planners run the risk of overestimating both costs (because, for example, certain implementation costs may be shared by several measures) and, more seriously, benefits (because, for example, the incremental effectiveness of additional measures may decline as more measures are implemented).

Some conservation measures may displease water users for reasons that cannot readily be translated into monetary terms. For example, residential users may be annoyed by the occasional need to double-flush a water-saving toilet, or by the inconvenience of adhering to sprinkling restrictions. The community as a whole may feel a sense of loss at the appearance of brown lawns and dying shrubbery during a summer drought that has led to sprinkling restrictions.

These examples illustrate the difficulties a forecaster is likely to encounter in trying to evaluate some types of adverse effects in monetary terms. Ultimately, most such effects must be treated as nonmonetary and must be analyzed descriptively rather than quantitatively. Little guidance is available for identifying and describing nonmonetary effects, but their potential presence should be recognized and their specific expected effects should be described whenever feasible.

## Dealing with Uncertainty

Careful, controlled measurements of effectiveness are available for only a few conservation measures. Even these measurements are frequently based on observations of very limited numbers of water users under severely restricted conditions of implementation. Such data, therefore, should be regarded as highly uncertain. One alternative to simply adopting the measurements of other researchers is to identify reasonable upper and lower bounds and the most likely level for effectiveness, based on the literature and on independent estimates (as required). In some cases, trial implementation of the measure for a small sample of water users may provide the basis for such estimates. Where available estimates of effectiveness span a very wide range, even a limited experiment may enable the forecaster to narrow the range of expected effectiveness.

To prepare estimates of conservation effectiveness, the forecaster must use a forecasting method capable of reflecting the impact of individual conservation measures that often apply to a single sector of use. In the case of urban water supply, the forecasting method must be disaggregate; that is, it must estimate future water use separately for the various sectors of water use (residential, industrial, commercial, and so on). Where the capacity of supply facilities is a critical issue, seasonal water uses must also be estimated separately. The forecast method must be able to forecast water use in all sectors in terms of various explanatory variables, including price. Per-capita forecasting techniques—although widely used in water resource planning—are incapable of producing reliable precise estimates of conservation effectiveness.

If potentially interactive measures have been successfully identified, grouped, and assessed in various combinations, the estimates of their effectiveness will be additive: if two or more measures or sets of measures are to be implemented concurrently, their combined effectiveness can be found by summing the individual estimates of effectiveness. Related estimates of implementation costs and of other beneficial and adverse effects can also be summed to obtain estimates for the combination of measures.

Estimates of effectiveness must be obtained throughout the planning period to give proper definition to the time pattern of benefits. This requires regular forecasts (every five years, for example) of water use with and without the conservation measures. The water-use reductions thus defined will be assumed to describe a smooth curve plotting effectiveness as a function of time.

## IMPACTS OF WATER QUALITY

## CLASSES OF WATER USE

For water supply, quality is perhaps the main criterion for distinguishing among different uses. A natural starting point for the water forecaster is to recognize that different uses of water require different quality characteristics. For example, dissolved-solids concentrations of 35,000 milligrams per liter (mg/l) are permissible in some industrial uses, whereas a maximum of 500 mg/l is recommended for public water supply.

Most supply/demand studies assume the supply of water at a given quality level and hence never deal with the issue when making supply/demand estimates or when considering resource-allocation problems. Further, quality is seldom defined in terms of the factors that might reduce it and thus reduce the supply available for particular uses. Often, studies simply mention that quality may vary, without attempting to identify quantitatively how it varies or what effect the variations may have on supply.

Water quality can be classified in at least three ways: by contaminants, by uses, and by treatment required to remove contaminants (Bovet 1973).

### Classification by Contaminant

The list of possible contaminants or impairments of water is extensive, and includes perceptible, microbiological, and inorganic constituents. Some items (such as pesticides, trace elements, and radiological substances) are little affected by common water-treatment processes; if any such item were found in a concentration exceeding the recommended maximum level, special treatment would be required. Items that can be removed by common treatment processes include the following:

1. **Perceptible Impairments**
   - Taste
   - Color
   - Odor
   - Temperature
2. **Microbiological Contaminants**
   - Bacteria count (Bacteria, especially fecal coliforms, have been used as indicators of the sanitary quality of water for nearly a century.)
3. **Inorganic Chemicals and Related Items**
   - Biochemical oxygen demand (BOD is a measure of how heavily taxed the natural assimilative capacity of a watercourse is. BOD varies inversely with the dissolved-oxygen (DO) content of a waterway.)
   - Suspended solids (These are items carried by, but not in solution with, a watercourse.)
   - Total dissolved solids (These consist mainly of elements and salts in solution.)
   - pH (pH is a measure of dissociated hydrogen ions in a liquid, and thus is an indicator of acidity/alkalinity.)
   - Hardness (Generally, hardness is a measure of the concentration of calcium carbonate ($CaCO_3$) in water, which is associated with scaling on pipes.)

## Classification by Use

Water-quality classification by use is an attempt to eliminate irrelevant water-quality parameters. Safe water-quality standards (or tolerances), expressed in terms of concentrations of foreign items, must be established for each use. This allows the water forecaster to distinguish between critical and noncritical parameters. For example, high mineral content is a critical problem in water used for domestic water supply; but in a recreational use such as swimming, even seawater can be tolerated. By ignoring parameters that, in a given water use, are noncritical, the forecaster can simplify the classification of water quality and at the same time make it more relevant to the matter at hand.

Different quality levels of water can also be compared more readily when they are related to a given water use.

User groups offer an efficient means to determine precisely which quality characteristics must be dealt with. A literature survey conducted by Ernst and Ernst (1973) for the Army Corps of Engineers identified the following eight classes of water use, based on quality considerations:

- Public drinking supply
- Aquatic (fish/shellfish) and wildlife propagation
- Recreation and aesthetics
- Industrial supply
- Agricultural use
- Water power (mainly hydroelectric)
- Navigation
- Disposal of industrial waste and sewage

The aquatic and wildlife propagation class refers to water used as natural habitat for animal species. Water use for waste disposal refers to the natural assimilative capacity of water courses. The other headings are, in general, self-explanatory.

Bovet (1973) provides a more detailed list of water uses as an alternative classification:

1. Public Water Supply
   - Residential or domestic use
   - Commercial use
   - Industrial use (of public supply)
   - Institutional use
   - Fire-fighting use
2. Industry (not using public supplies)
   - Boiler makeup use
   - Processing use
   - Product use
   - Cooling use
   - Sanitary use
   - Fire-fighting use
3. Recreation
   - Water-contact activities (swimming, diving, skiing)
   - Water-based activities (boating, fishing)

- Water-related activities (picnicking, camping, hunting)
- Use of park facilities (swimming pool use, drinking, cooking)

4. Plants and Animals
   - Phytoplankton, water plants
   - Zooplankton
   - Macroinvertebrates, fish, reptiles
   - Waterfowl, other aquatic animal life
   - Wildlife
5. Waste Producers
6. Aesthetic Enjoyment
7. Regional Planning

## Classification by Treatment Cost

Water-quality classification by treatment cost constitutes another substantial simplification. Under this system, water quality is measured by the cost involved in bringing it to levels acceptable for given water uses. Required data include tolerances or standards for various water uses, effects of water treatment, percentage removal of critical parameters achieved by such treatment, and cost of such treatment. Most water treatment processes are capable of removing more than one parameter, although sometimes not by the same percent. By adding appropriate treatment processes, utilities can raise water quality to any desired level of purity.

Table 2-9 summarizes the physical/biological/chemical treatment processes most commonly used to remove or reduce specific contaminants found in a water supply. Cost data vary as a result of a number of factors and can be obtained from appropriate reference books on treatment processes.

## WATER-QUALITY INDICES

In order to avoid the problems associated with direct measurements of individual quality factors and corresponding tolerances for each class of water use, planners can deal explicitly with quality through the use of an index number. This kind of quality function permits several quality factors to be included in a single dimensionless number. Indices can be computed by recording actual observations, soliciting opinions from knowledgeable people about how good or bad a body of water is, or using a combination of both methods.

**Table 2-9. Commonly Used Water Treatments and Principal Impairments Affected.**

| Treatment | How It Works | Impairment(s) Corrected |
|---|---|---|
| Screening | Water flows through porous grates | Large floating items |
| Grit chambers, comminutors | Shredding and separating devices | Gross solid pollutants |
| Coagulation, flocculation | Chemicals (alum, hydrated lime) cause agglomeration on surface waters | Turbidity, color, taste, odor, phosphorus, pH (coagulant aid especially effective) |
| Sedimentation | Removal of solid particles by gravitational settling | Suspended solids, turbidity, some BOD effect |
| Slow sand filter (not often used) | Separates substances by combination of straining, absorption, and flocculation | Bacteria, turbidity, color |
| Rapid sand filter | Sand/gravel medium removes nonsettleable floc and impurities remaining from coagulation | Bacteria, turbidity |
| Trickling filter | Waste effluent is sprayed over rock bed on which microorganisms grow and feed on organic matter | BOD, suspended solids, bacteria |
| Activated sludge | Process of aerating wastewater so microbiological waste metabolism will be faster | BOD, suspended solids, bacteria |
| Stabilization | Cachement for impounding water until organic wastes stabilize and aerobic decomposition occurs | BOD, suspended solids, bacteria |
| Disinfection (mainly chlorine) | Hydrochlorous acid forms, enabling chlorine to destroy bacteria cells' enzymatic processes | Bacteria (to lesser extent: odor, corrosion, BOD) |
| Activated carbon | Removes organic contaminants by absorption | Taste, odors |
| Ion exchange | Exchange resin removes certain metal ions in exchange for sodium | Hardness, dissolved solids, pH adjustment; also chlorine removal (reverse osmosis, electrodialysis also affect these parameters) |

SOURCE: Ernst and Ernst (1973), p. III-8.

An index is a number (usually dimensionless) whose value expresses a measure or estimate of the relative magnitude of some condition, such as the pollution level of a body of water; the index is a shorthand expression for a complex combination of several factors. For example, the overall system of water supply, pollution control, and human use of water resources consists of a complex interactive assemblage of meteorological, hydrological, industrial, chemical, physical, biological, engineering, sociological, economic, political, and institutional factors. The interrelationships among these factors are in most cases not clearly understood, and the necessary data for a rigorous analysis are only available at great expense (if at all). It is simply impractical to take every single factor into consideration.

An index's usefulness depends on its general acceptability as part of a system based on a consistent and careful method, rather than on any rigorous scientific justification. Each index should be carefully defined and computed so that it expresses the relevant condition as closely as practicable. No single number, however, can contain all relevant information about a complex situation.

Two related terms are *parameter* and *indicator*. A *parameter* is simply a measurable characteristic of the system under study; temperatures, concentrations, and flow rates are all parameters. Part of the technique of designing indices lies in selecting parameters that can be efficiently measured or estimated and can meaningfully characterize important states of the system. An index is usually an arithmetic combination of parameter values. An *indicator* is a parameter that correlates reasonably closely with an important condition that is less easily measured. The concentration of fecal coliform bacteria, for example, is used not only as an important characteristic of a body of water, but also as an indicator of the presence of other organisms and organic matter derived from municipal sewage systems.

Once defined, understood, and accepted, index values can quickly be grasped and compared in many cases where assimilating and comparing a complicated set of data would be unacceptably time-consuming and confusing. There will, of course, be many situations in which the rough approximation represented by an index will not suffice. In these instances, only a detailed analysis of the basic data will be satisfactorily rigorous.

Some practitioners insist that an index must be dimensionless, but indices often carry a dimension for added intuitive appeal. For example,

the Consumer Price Index, which is widely used in economic discussions, is dimensionless, while the Gross National Product index
carries the dimension of dollars.

Water-quality indices that have been proposed for use in water-
planning studies include the Bovet index, Mitre Corporation's PDI,
Syracuse University's PI, and the National Science Foundation's WQI.
The characteristics of these indices are discussed in the following
paragraphs.

## Bovet Water Quality Index

Bovet (1973) suggests using the cost of bringing water quality to
desirable levels as one type of quality index. Parameters that are
critical in a given water use can be combined into groups, thus simplifying
the classification of water quality. For a given water use, costs of all
required processes (expressed for a specified plant size, in dollars of
a specified year) are simply aggregated into a total amount. Classifying
water of any description can thus be reduced to measuring a single
dimension per water use. Water sources of widely differing quality
can readily be compared. And for each water use, the aggregate
treatment cost—upon identification with an appropriate cost bracket—
can be used as a water-quality index.

## Mitre Corporation's PDI

Developed for the Environmental Protection Agency's Office of Water
Programs, *PDI* stands for Prevalence–Duration–Intensity (Mitre Corporation 1972). Prevalence ($P$) is the number of miles (in the form of
stream length or waterfront) of uniformly polluted water in the pollution
zone. Duration ($D$) indicates the fraction (in quarters) of each year
during which the pollution zone exhibits the associated value of pollution
intensity. Values are assigned to $D$ as follows:

$$D = 0.4 \text{ for one quarter}$$
$$D = 0.6 \text{ for two quarters}$$
$$D = 0.8 \text{ for three quarters}$$
$$D = 1.0 \text{ for the entire year}$$

Intensity ($I$) indicates the severity of pollution in the pollution zone
during the portion of the year represented by the $D$ value assigned.

Mitre defines the $I$ index as consisting of the following components: $I_1$ (ecological) is the index of how damaging to life the pollution is; $I_2$ (utilitarian) is the index of how disruptive to normal uses the pollution is; and $I_3$ (aesthetics) is the index of visual and olfactory unpleasantness. Permissible values, which translate opinions into numerical responses, are specified for each component; thus, the opinions can readily be rendered in numerical form for each $I(i)$. Therefore, $I = I_1 + I_2 + I_3$. The three components of $I$ are described as follows:

1. Ecological
    $I_1 = 0.1$—conditions that threaten to produce stress on life forms
    $I_1 = 0.2$—conditions that produce stress on indigenous life forms
    $I_1 = 0.3$—conditions that reduce productivity of indigenous life forms
    $I_1 = 0.4$—conditions that inhibit normal life processes or threaten elimination of indigenous life forms
    $I_1 = 0.5$—conditions that eliminate one or more indigenous life forms
2. Utilitarian
    $I_2 = 0.1$—conditions that require above-normal costs to support legally defined uses
    $I_2 = 0.2$—conditions that intermittently inhibit realization of some desired and practical use or necessitate use of an alternate source
    $I_2 = 0.3$—conditions that frequently or continuously prevent the realization of desired and practical uses or cause physical damage to facilities
3. Aesthetic
    $I_3 = 0.1$—visually unpleasant
    $I_3 = 0.2$—visually unpleasant, with accompanying unpleasant tastes or odors

As noted, $P$ is simply measured in miles, while $D$ is a time measure; accordingly, both are data observations. On the other hand, $I$ is subjectively determined by experts' opinions. In the Mitre study, regional offices of the Environmental Protection Agency provided estimates of PDI for specified water bodies.

For example, suppose that a water body 10 miles long is known to experience pollution for nine months of the year. These objective observations are translated into the values $P = 10$ miles and $D = 0.8$ (dimensionless) in Mitre's formulation. Now assume that subjective determination has provided observations that are transformed by Mitre's scales into the following values: $I_1 = 0.5$ ("conditions that eliminate one or more indigenous life forms"); $I_2 = 0.2$ ("conditions that intermittently inhibit realization of some desired and practical use or necessitate use of an alternate source"); and $I_3 = 0.2$ ("visually unpleasant, with accompanying unpleasant tastes or odors"). By addition, $I = I_1 + I_2 + I_3 = 0.9$; so the complete index for the hypothetical water reach in question is $P \times D \times I = 10 \times 0.8 \times 0.9 = 7.2$.

Although PDI can be considered a first-hand operational tool for the water planner, it has several weaknesses. The $D$ and $I$ components are defined as numbers between 0 and 1; so when the index is formed as the product $P \times D \times I$, it is actually a weighted miles-of-length measure—higher values of which indicate a more severe pollution problem. This mileage-dominance trait could lead to a situation in which high values of $D$ and $I$ applied to a low value of $P$ produce a lower PDI value than do low values of $D$ and $I$ applied to a high value of $P$. This might lead some people to judge the former instance as the more severe problem. The index must therefore be used carefully; its meaning is probably least ambiguous when the index is applied to reaches of similar length.

A second weakness is that PDI does not explicitly account for pollutant types. Even though data on water-quality conditions exists, there is no assurance that Environmental Protection Agency personnel have used them in estimating $I$. And even if such information has been used, the PDI index gives no indication that the data have been applied in a uniform, systematic, and replicable manner. Thus, for example, when an expert assigns $I_1 = 0.2$ ("conditions that produce stress on indigenous life forms") to a particular reach of water, the action represents a determination—based on an internal synthesis by the expert of knowledge about pH, turbidity, BOD, and so on—that, of the five conditions Mitre allows to be assigned to $I_2$, a value corresponding to $I_2 = 0.2$ most accurately represents the pollution intensity on the watercourse being evaluated. The quality factors themselves do not appear in PDI.

### Syracuse University Civil Engineering Department's PI

Syracuse's PI (Pollution Index) is based on explicit consideration of fourteen monitored water-quality factors: temperature, color, turbidity, bacteria, total solids, suspended solids, total nitrate, alkalinity, hardness, chloride, iron, manganese, sulfate, and dissolved oxygen. For a specified *j*th water use, two observations are needed: the current measured level, $C_{ij}$, of each *i*th impairment; and the recommended tolerance, $L_{ij}$, by use *j* of each impairment *i*. The dimensionless ratio, $R_{ij} = C_{ij}/L_{ij}$, thus indicates how critical the level of impairment *i* is in use *j*. A value of $R_{ij}$ greater than 1 indicates that treatment is needed, because the monitored level exceeds the tolerable level.

The $PI_j$ for use *j* is formed by first computing $R_{ij}$ for the fourteen quality parameters listed. Then the mean $\bar{R}_j$ of these values is calculated, as an indication of the average impairment concentration. The maximum of the $R_{ij}$ values (denoted max $R_{ij}$) is used along with $\bar{R}_j$ in computing $PI_j$, since the necessity of water treatment prior to a specified use is often determined by the maximum of the $R_{ij}$ values, rather than by the average value. The pollution index is obtained by means of the following equation:

$$PI_j = \sqrt{[\bar{R}_j^2 + (\max R_{ij})^2]/2} \qquad (2.5)$$

The PI can be computed for an entire region by calculating the weighted average of the use *j* values in a region. In addition, $PI_j$ values can be calculated for groups of water users. The Syracuse authors suggest three groups: human contact use (drinking, swimming, beverage manufacture, and so on); indirect contact use (fishing, agricultural use, and so on); and remote contact use (industrial cooling, aesthetics, navigation, and so on).

The PI formulation affords a greater degree of technical precision than does the PDI formulation because it is totally objective—thus averting the major criticism leveled at the PDI index. On the other hand, this objectivity depends on the availability of the requisite quality-factor observations (assuming that accurate means of monitoring have been employed). In addition, a pragmatic water planner might bemoan the fact that PI ignores the size of and time element relevant to the water body being examined (in short the *P* and the *D* of PDI). PI is essentially a spot analysis.

### National Sanitation Foundation's WQI

The Water Quality Index (WQI) combines the subjective nature of PDI's *I* component and the objectivity of PI. It is constructed by collecting water experts' opinions about the severity of specified individual impairments in a water body and then synthesizing the responses into a single representation. Each expert was first asked (via questionnaire) to state which items from a list of water-quality factors were most important to review in water analyses. From the responses, a list of nine "most important" parameters was chosen for inclusion in the ultimate WQI: dissolved oxygen, fecal coliform count, pH, BOD, nitrates, phosphates, temperature, turbidity, and total solids.

In a subsequent questionnaire, respondents were asked to assign values for the variations in water-quality level produced by different strengths of the nine selected parameters. This was done by sketching a representative curve on a graph whose coordinates were water quality (as dimensionless numbers between 0 and 100) and parameter strength (measured in typical units for the parameter). The average of all the respondents' sketched curves for each factor was then deduced, to be used as a transformation function for translating any given parameter concentration into a quality-level number (symbolically, $q_i = f(m_i)$, where $q_i$ is the quality number and $m_i$ is the measured parameter concentration). Finally, the respondents' comments about the relative importance of the parameters were used to construct parameter weights, $w_i$, so that the nine individual quality-level values, $q_i$, could be combined into one index, as follows:

$$WQI = \sum_{i=1}^{9} w_i q_i \tag{2.6}$$

The procedure for applying equation 2.6 would be first to obtain concentration measurements, $m_i$, on each of the nine parameters ($i = 1, \ldots, 9$) for a water course, then to translate these observations into quality values by means of the $q_i = f(m_i)$ functions, and finally to sum these values, weighted by the $w_i$. The result is clearly a quasi-objective index. WQI is proposed as a general, overall index; there is no attempt (as there is with PI) to propose user-specific $WQI$ values.

WQI shares PI's weakness of not incorporating $P$ (space) and $D$ (time) elements. And like PDI, WQI has subjective content that some might criticize. The subjective content of WQI, however, is interspersed with more empirical data. In water planning, WQI's subjective content may be an asset if it happens that the expert opinion determinations of $w_i$ and $q_i$ permit WQI to indicate the severity of an acknowledged pollution problem more emphatically than does the purely objective determination of $R_{ij}$ in PI.

## IMPACTS OF WATER RECLAMATION

The use of reclaimed water has become an increasingly important quality issue in the forecasting process. In chapter 1, a distinction was drawn between *reuse of water by industry* (which is defined as direct reuse of water, with no or limited treatment, at the same general location or for the same purpose) and *reclaimed water* (which is defined as water that has been treated in a reclamation plant and is usually transported to another location for reuse). Baumann (1981) expands the definition of *reuse* to include the collection and treatment of sewage and the use of the effluent for irrigation, recreation, industry, or general municipal use (upon direct return through an intervening body of water or aquifer).

Renovation and reuse of municipal water are not new concepts. On average, approximately one-third of the population in the United States relies on municipal withdrawals from streams containing 1 gallon of previously used water for every 30 gallons withdrawn (Koenig 1966). In some instances, the ratio of previously used water is as high as 1:5 (Baumann 1981).

In response to increases in demand for water, water planners have traditionally chosen to increase the available supply. Except under emergency conditions, such as drought, alternatives that would reduce demand were rarely considered or adopted. As the process of urbanization continues, the size of demand for water within a relatively small area will add to the pressure for greater efficiency in water use, as well as creating intense competition over rights to the available supply. Table 2-10 summarizes the alternatives available in balancing supply and demand for urban water.

### Supply Alternatives

The major source of municipal water supply in the United States—accounting for 75 percent of total capacity—is diverted rivers and streams. Use of surface-water flows is projected to increase as a proportion of total water use. As urban areas have grown, however, the streams nearest to the cities have been developed, and opportunities for future diversions are becoming more scarce.

In regions of the United States where water shortages are most common, interbasin transfers of water are required. But such transfers are often expensive, with the economic, political, environmental, and technological difficulties increasing as the distance of the transfer

**Table 2-10. Alternatives for Balancing Supply and Demand for Municipal Water Supply.**

| Do Nothing | Modify Supply | Modify Demand |
|---|---|---|
| Accept shortage<br>• unplanned rationing | Increase supply<br>• divert new streams<br>• provide increased storage<br>• use groundwater | Restrictions<br><br>Price elasticity<br>• peak pricing (peak summer pricing)<br>• marginal-cost pricing |
| | Increase efficiency<br>• reduce reservoir evaporation<br>• eliminate leaks<br>• increase runoff<br>• reduce evaporation | Meters<br><br>Educational campaign<br>• emphasize water<br>• use conservation |
| | Weather modification<br><br>Desalinization<br><br>Renovated wastewater<br>• nonpotable uses<br>• potable uses | Technological innovations and applications (e.g., changes from water cooling to air cooling) |

SOURCE: Baumann (1981), p. 274.

increases. The area where the water originates (the donor region) is usually rural and often has a higher incidence of individual well users. These users tend to regard the water as belonging to the region and to view its transfer as having been necessitated by the unreasonable water-use habits of urban water users.

Use of groundwater, which currently accounts for 25 percent of total municipal supply, is projected to decrease slightly as a proportion of total water use. Sources capable of sustaining high withdrawal rates are limited in distribution, and—while groundwater is the predominant source of self-supplied individual users—most major cities tapping groundwater supplies use it as a supplement to their primary source of supply.

Desalinization and weather modification may, in selected circumstances, serve to augment conventional water supplies. Although highly variable in its effects, weather modification in some instances has considerable potential, such as in increasing snow pack and the subsequent spring runoff during years when winter reservoir storage is low. Desalinization is comparatively expensive, requires large amounts of energy, and is most promising on a small scale in unique situations.

Although efforts to reduce seepage and evaporation have met with little success, reduction of water loss by identifying previously undetected leaks can be substantial. As much as 4 percent of all water withdrawn may be unaccounted for because of leakage.

### Demand Alternatives

Although other alternatives are available and practicable, water planners traditionally have seen supply as the variable in the supply/demand equation. Price has been shown to be a significant variable, but it is usually disregarded as a method of controlling demand. Economists are quick to point out that marginal costs should equal price; the decreasing-block pricing system common to most cities, however, encourages a high level of water use and prices the last gallon of flow—which normally is the most costly—at the lowest price. Increasing-block rates, peak-summer pricing, and annual rate changes based on the supply in storage have all been proposed as methods of reflecting marginal cost in the pricing of water. Water pricing is discussed in detail in chapter 3.

Rationing water and restricting its uses have been used as a management tool predominantly during crises. While the cost of these measures to consumers during periods of water restriction has been low, they are not usually regarded as a potential method of choice in planning for water supply. Moreover, little basis exists for evaluating the long-term effects of encouraging the public to reduce water consumption.

## FORECAST APPLICATIONS

Consideration of the topic of reuse should start with consideration of the concept of an integrated system—integrated in both the management and the distribution of water, so that either treated effluent or potable water can be supplied for nonpotable uses (Dworkin 1981). With this control, the water planner's judgment about the state of the system can be enlisted to determine whether to produce and distribute treated effluent for certain uses or to use only the potable supply in all cases. This would eliminate unnecessary costs incurred in producing and distributing effluent during periods when unused capacity exists in the potable water system.

In general, the relationship between reuse and potable water storage provides a guide to using treated effluent. Reuse systems typically cost less to build but are more expensive to operate than systems designed to divert and store potable water. On the other hand, potable-water systems are expensive, but have low operating costs—a feature suggesting that reuse systems should be used only when potable water from storage is unavailable to meet the demands. In this way, reuse would function as a source of peak supply, while potable water would provide the cost-effective base-load capacity. There are limitations, however. Treated effluent would only be supplied to users previously identified and equipped with a separate distribution system for non-potable uses. Still, these limitations only serve as constraints on the realization of the system's full potential.

### Benefits of Reuse

Reuse, whatever its end application may be, provides a source of water that can delay or eliminate the need for conventional additions to supply. The presence of reuse as a standby source unaffected by periods of low flow can increase system yield and provide planning flexibility in the following ways: as a substitute for the high levels of water-supply assurance required in municipal systems; as a means of mobilizing any excess supply in the system; and as a means of shortening the planning cycle to allow pragmatic evaluation of changes in demand in lieu of long-term projections.

**Supply Assurance**   The yield of a water-supply system based on potable-water storage is usually expressed as a quantity of water available or assured at least 98 percent of the time. To achieve this level of assurance (or reliability), some storage must be provided that will be required less than 5 percent of the time. An inverse relationship exists between assurance and yield, such that yield increases as assurance declines. When the levels of assurance in a system are relaxed, the yields withdrawn from the system can be increased, with reuse furnishing the additional water necessary to maintain the desired levels of reliability.

**Mobilizing Excess Supply**   Because the yield of urban systems is always calculated to provide high levels of reliability, periodic rationing or restriction of water causes little damage. Even a shortage occurring less than 5 percent of the time, however, is not acceptable to engineers. Engineering and reference handbooks urge conservative calculations, while social scientists claim that yields are often (if not always) understated. As an alternative, water reuse can provide a standby source that allows existing facilities to be used until pragmatic evaluations can be made of the response of the physical system to future demands.

**Shortening the Planning Cycle**   The long time needed to develop new water sources requires long-term projections of the future demand for water. In the past, planners were little concerned if future demand was overstated and resulted in temporary oversupply. The rapid growth in water use could be counted on to render any excess capacity useful in a short time. The low rate of births in recent years and the replacement of single-family houses with apartments and cluster homes, however, indicate that the days of unrelenting rapid growth in demand are over. Excess capacity added now, under lower rates of growth, will be utilized more slowly; and consequently the investment in idle capacity will extend over a longer period.

### Factors Affecting Feasibility

The present method of evaluating the economic efficiency of reuse by comparing the costs of reuse and freshwater-use alternatives is misleading. The most important consideration in any economic analysis is the cost at the margin. If two systems of supplying water have the same average costs, and if reuse is considered as a supplement to both, an average cost analysis of the operations of both systems would produce the same findings. But if no excess capacity exists, and if the alternative is to develop conventional supplies, providing for reuse might be a more efficient economic solution. Calculating the costs of the final unit of water delivered in a system with and without reuse should be the criterion used in any economic evaluation.

Evaluation of reuse should not consist exclusively of comparing the two alternatives, reuse and conventional supply. The analysis should be extended to include a determination of the best method of operating the reuse facility to provide optimal long-term solutions to the problem of meeting water demands. In such an analysis, reuse should be considered as one of the sources of supply, along with surface water and groundwater, and the demands for water should be critically appraised. The water planner should be encouraged to manipulate the four components—groundwater, surface water, reuse, and demand—to achieve the most efficient system.

## INTEGRATED APPROACH TO DEMAND FORECASTING

The preceding discussion has shown that the water planner must contend with myriad changing issues during the forecasting process. These issues include future growth of the service area, potential competition for service territory by other suppliers, management of peak demands on the system, effects of conservation programs, and water-quality constraints for classes of users.

The changing issues faced by water planners have supported the development and use of advanced management techniques and demand-forecasting models. Although appalling to some engineers who are comfortable with the relatively precise and immutable principles of hydraulics, this integrated approach to demand forecasting has been mandated by changes in the analytical environment; its use thus represents a necessary adaptation by planners to changing data bases and information requirements (DeKay 1985).

## SEATTLE WATER DEPARTMENT'S EXPERIENCE

Over the last seventy-five years, the Seattle Water Department's demand forecasts have evolved from intuitive assessments to sophisticated computer models that prepare geographically disaggregated, long-term forecasts for planning facilities and customer-class-disaggregated short-term forecasts for rate-making and financial planning. This section presents a chronology of this evolution, a discussion of the issues that spawned it, and an analysis of the data bases that made it possible (DeKay 1985).

The Seattle Water Department (SWD) is a regional water supplier for the entire city of Seattle and for some outlying areas. It serves approximately 1.1 million people—approximately half of them as direct customers, and the remainder through wholesaling of water to thirty-two local communities and water-supply districts. In 1982, SWD sold 47.6 billion gallons of water, which amounts to roughly 120 gpcd. About 54 percent of the water sold by the Seattle Water Department is used in single-family residences; multifamily residences, governmental users, commercial users, and industrial users account for 11 to 12 percent each.

Table 2-11 presents an outline of the evolution of the Seattle Water Department's forecasting activities for 1940 to 1983. As shown in table 2-10, emerging issues motivate forecasts, and data bases place constraints on the range of techniques that may be used in customized forecasts.

### Early Forecasting Models (1940–73)

Planning efforts during the period from 1940 to 1973 used per-capita methods for estimating water consumption. In 1940, a water use of 127 gpcd was reported for the entire system. Later efforts undertook more precise analyses by providing per-capita estimates for geographical subareas and separate user types. In a 1968 study, the service area was divided into thirteen geographic subdivisions, and population and employment projections for each area were made on the basis of projected land use. Water-use forecasts were developed by determining per-unit water use for different users and multiplying these figures by the estimated number of users in each geographic area.

Total water use was disaggregated into three components: industrial, commercial, and miscellaneous. Water use per industrial employee was computed as total water sales in an industrial area divided by employment in that area over a seven-year period. The resulting unit factor of 185 gpcd was assumed for all industrial employees in the service area. Water use per commercial employee was computed as total water use in the central business district divided by employment in that area. This yielded a use rate of 70 gpcd, which was then applied to all geographic regions. Water use outside the central business district and the industrial area was divided by the figure for total population to obtain a unit factor of 90 gpcd for all other water uses. When the three unit factors were aggregated, total water use was estimated to be about 135 gpcd for 1968.

A 1970 rate study designed to forecast revenues rather than water use retained the idea of geographic disaggregation by analyzing data from several geographic subareas, but it abandoned analysis on a user-

**Table 2-11. Evolution of Demand Forecasting Techniques and Data Bases for the Seattle Water Department, 1940–1984.**

| Time Period | Issues | Techniques | Data Bases |
|---|---|---|---|
| 1940–1950 | | Per-capita methods | System's total diversions |
| 1950–1965 | Planning size and timing of major facilities | Per-capita methods | System's total diversions |
| 1965–1970 | Long-range planning | Unit-use coefficient method | Billing data from three geographic areas: CBD, industrial, all other |
| 1970–1974 | Revenue forecasts for rate studies | Judgmental methods | Three geographic areas: in city, outside city, wholesale |
| 1973–1975 | Regional water planning | Multivariate requirements models<br>Demand model methods<br>Scenarios<br>Sensitivity analysis | Cross-sectional billing data for thirteen residential areas<br>Retail sales for four sectors for one year<br>Time-series for system's total diversions |
| 1976–1979 | Medium- and long-range facilities planning<br>Hydraulic modeling<br>Conservation plan analysis | Demand model methods<br>Sensitivity analysis<br>Scenarios<br>Confidence intervals<br>Per-capita methods | Three years of monthly sales data for five sectors<br>Cross-sectional data for sixty-three residential areas |
| 1979–1981 | Short-term revenue forecasts for rate setting<br>Conservation plan analysis | Time-series analysis combined with multivariate regression<br>Confidence intervals | Five years of monthly sales data for eleven sectors |
| 1982–1984 | Medium- and long-range facilities planning<br>Revenue forecasts for rate setting<br>Conservation plan assessment<br>Analysis of costs of shortages<br>Risk analysis | Demand model methods<br>Scenarios<br>Confidence intervals<br>Judgmental estimates of long-term price elasticities<br>Monte Carlo studies<br>Per-capita methods | Five to eight years of wet season, dry season data for sixty revenue classes<br>Cross-sectional data by census tract for all revenue classes |

SOURCE: Reprinted from *Journal of the American Water Works Association* 77(10), by permission; copyright © 1985, American Water Works Association.

sector basis. The study assumed that, over the next six years, revenues would grow at 0 percent per year for the areas served inside the city, at 4 percent per year for the retail areas served adjacent to the city, and at 8 percent per year for the areas served by the purveyors. Historical data on population and on the number of meters accompanied these forecasts, but the growth rates themselves were judgmentally determined from historical trends.

From the perspective of future forecasting efforts, the most important contribution of this study was its recommendation that billing data be coded to allow identification of water use by different types of users. More than thirty types of users were defined for both in-city and out-of-city customers, allowing identification of more than sixty revenue classes. This recommendation, made possible with the advent of computer technology, allowed all future studies to group classes of users into sectors according to rate class, customer class, or other relevant category for analysis.

The next four studies are characterized by their progressively sophisticated forecasting methods; each is discussed in detail on the pages that follow.

### River Basin Study (1973)
In 1973, the Pacific Northwest River Basin's Coordination Committee (RIBCO) sponsored a series of studies of water resources in the Seattle region, analyzing total system flows with time-series data in a multivariate model. Cross-sectional data were used to analyze residential use, and unit-use coefficient methods were used to analyze nonresidential water use. Data on total system diversions were available for the period from 1950 to 1972, allowing regression analysis of total system water use from May to September of each year against population, precipitation, and air temperature. For the first time, water-sales data were available (albeit only for the year 1972) by category of user. From the sixty or more revenue classes identified in the 1970 rate study, five major sectors were defined: single-family residential, multi-family residential, commercial, industrial, and public and miscellaneous users.

In the RIBCO study, several scenarios of water use were prepared in an attempt to identify a range of forecasts. High, medium, and low forecasts were prepared on the basis of different assumptions about population growth, future income, and water productivity in industrial applications. Forecasts were prepared for ten-year intervals to the year 2020. The sensitivity of the forecasts to different assumptions was noted.

The improved geographic, sectoral, and seasonal disaggregation allowed by the new data base led to a better understanding of water use in the Seattle Water Department's use area. This study marked the first attempt at regression analysis of time-series and cross-sectional data for forecasting. It also marked the first such use of scenario analysis (preparing several forecasts that incorporate different combinations of assumptions) and sensitivity analysis (investigating the changes in a forecast when a single variable is changed by a specified amount).

### Seattle Metropolitan Water Supply Study (1977)

This study addressed long-term source planning and medium-term system improvements. Consequently, forecasts were required for the system as a whole, for individual geographic areas (for use in a hydraulic model), and for different sectors. In the 1977 study, data for about thirty different types of users during the 1972–76 period were accumulated. For the first time, analysis of time-series data by sector was practicable, because monthly data was available for many sectors.

Users were aggregated into five sectors for analysis: single-family residential, multifamily residential, commercial, industrial, and government and miscellaneous. The emerging availability of a detailed data base promoted a more sophisticated analysis of water use.

Using the model, the Seattle Water Department developed high, medium, and low forecasts based on different assumptions about regional development, real income growth, real price changes, per-employee trends, and conservation behavior. To determine which variable most influenced the forecasts of future water use, sensitivity analyses were performed on each variable in the model. Regional growth variables— such as the number of households and the number of employees— were clearly the most critical.

Statistical variations in water use, as measured in the regression equations, were used to estimate statistical confidence intervals at an assumed reliability level of 90 percent. The short time period on which the empirical analysis was based limited the reliability of these estimates, but the combined intervals (predicted from the sectoral equations) matched the variability of total water use closely enough over an extended time period to be considered useful.

A different forecasting method was used for planning source development. These long-term (fifty-year) forecasts were based on per-capita usage, as estimated from the computer model, multiplied by long-term population forecasts. Less effort was spent on the long term because of the unavoidable uncertainty in forecasting so far into the future; in any case, revisions of such forecasts to reflect new information can be introduced before irreversible decisions must be made.

### Rate Study (1980)

The Seattle Water Department and its consultants prepared water-sales and water-revenue forecasts for eleven categories of users, including wholesale, single-family residential, multifamily residential, commercial, industrial, and governmental. The retail categories were further disaggregated into retail customers inside the city and retail customers outside the city. Simple trending was used to estimate the number of customers.

The 1980 model analyzed water use on a sectoral-total basis, rather than on a per-user basis as previous studies had; the latter approach had required per-user estimates to develop geographically disaggregated forecasts. The 1980 model also used more sophisticated estimating methods and functional forms than did previous studies, which had relied on ordinary least-squares regression techniques to estimate water-use forecasting equations. In the 1980 model, analysts estimated statistical confidence intervals, allowing the uncertainty implicit in all forecasts to be explicitly considered in management decisions.

The increasing focus on financial management heightened interest in financial planning and, ultimately, in the forecasting of sales and revenues. Earlier rate studies (in 1970 and 1974) had used rather crude methods of estimating revenues. The desire for more reliable forecasts led planners to devote more resources to forecasting future sales. Since annual revenue forecasts were needed for a five- to seven-year time horizon, short-term forecasting methods were adopted. By this time, monthly data over a five-year period were available for all types of users. The expanded data base was sufficient to support a sophisticated combination of regression and time-series analyses to examine water-use patterns.

## Comprehensive Plan (1983)

The most recent forecasting efforts of the Seattle Water Department have been designed to provide coordinated development of consistent long- and short-range forecasts and to establish links between rate, sales, and facilities plans. As time passes, additional observations on water use are recorded, fortifying the data base and expanding the information available on water-use patterns. A minor addition to water-billing records has contributed significantly to their usefulness as a data source for forecasting. Early in 1983, census-tract designations were added to all accounts, allowing all billing data to be geographically aggregated by census tract. These data are now being used in developing long-term forecasting models of residential use.

The Seattle Water Department has prepared forecasting models for horizons of thirty and fifty years covering plant expansions, system improvements, and source development. For the thirty-year horizon model, four sectors of users were defined: single-family households, multifamily households, business users, and governmental users. Cross-sectional data by census tract, derived from the 1980 billing records, were used to estimate regression equations; the parameters of the forecasting equations were taken from these equations. The cross-sectional data provided independent variables having the smallest possible errors in measurement, and thus the greatest statistical efficiency within inherent constraints. Cross-sectional data also provided a long-term response to changes, because the data across observations are assumed to be adjusted to long-term equilibrium.

## CONCLUSIONS

One theme emerging from the Seattle Water Department's experience is the value of disaggregated models. The most common disaggregations are by user sector (residential, commercial, industrial, and so on), by time of year (winter versus summer, seasonal use versus nonseasonal use), and by geographic subarea. Sectoral disaggregation can range from a two-way split (usually into municipal and industrial sectors) to very detailed divisions within the traditional sectors.

Jones et al. (1984) cite two main purposes of disaggregation. First, disaggregation allows each individual sector to be described and forecast in terms of explanatory variables that uniquely relate to or affect it. Second, disaggregation produces detailed forecasts that may be needed

for planning or for evaluating specific strategies such as water conservation. Where sufficient data of adequate quality are available,
disaggregated models produce more accurate and useful forecasts than
do simpler, aggregated models.

Collecting data on a disaggregated basis also increases the probability
that, if billing procedures change, any negative impact of the change
on the quality of the data can be mitigated by reaggregating the data
into different classes of users or different time periods. Of course,
data collection is expensive, and the marginal efforts expended should
be equated to the marginal benefits anticipated from the additional
data collected.

The most valuable and accessible data gathered in the Seattle water
studies were obtained from computerized billings that could be processed
into the desired formats. Although all major water systems use computerized billing, not all such billings are easily usable for generating
data bases. Since a great deal of Seattle's data were collected monthly,
each month's information could be used to investigate short-term
patterns or could be aggregated to forecast seasonal use. Eventually,
when sufficient years passed, the data could be aggregated into annual
divisions to identify long-term trends.

The progress in forecasting achieved by the Seattle Water Department
over the past several decades has been paced by the evolution of
available data bases and data-processing equipment. The first of several
milestones in database development was the utilization of a computerized
billing system capable of generating periodic consumption reports. A
second key advance came in 1970 with the addition of customer-type
codes to accounting records, allowing segregation of water use by
sector. The designation of many more classes of users than were then
needed made flexible definitions of customer classes possible at a later
date, when more data were available and when forecasting needs and
techniques had developed. Adding census-tract codes to the billing
records in 1983 was a third key database improvement; it allowed
geographic disaggregation for cross-sectional data analysis.

The Seattle Water Department's billing system is now capable of
providing consumption data by geographic area and user type at a
level of disaggregation beyond what is currently used for analytic
purposes. Although this leads to additional expense and involves continuing efforts to maintain and update the data base, the history of
forecasting activities illustrates the usefulness of such efforts. Data-

processing costs have fallen dramatically in the past decade, making detailed analysis economically feasible if the data are available. Current forecasting activities analyze water use in sixty categories, allowing identification of trends, price elasticities, and the effects of weather on different user types. This contributes to increased confidence in water-use predictions and increased understanding of water-use patterns.

The Seattle Water Department's ability to implement these tactics can be traced to its decision, made more than ten years ago, to collect data on a more disaggregated basis than was used for analysis at that time. The foresight of the decision not only enabled planners to perform more detailed analyses and to apply specialized techniques, it also enabled them to preserve the integrity of the time-series data.

A significant feature of the Seattle experience is the coordinated use of cross-sectional and time-series data. Studies instituted in 1977, for example, used time-series data to assess climatic influences on water demand. Since these time-series data lacked substantial price variation or a range of values for other strategic variables—such as water pressure or property value—cross-sectional data were used to analyze the effects of changes in these explanatory variables. Coefficients from the two types of models were synthesized in a highly disaggregated model. Distribution needs in 120 zones of the service area, based on thirteen water-demand or multivariate relationships, were evaluated over a ten- to twenty-year period so that weaknesses and areas of low reliability in the hydraulic system could be identified.

Another point illustrated by the evolution of the Seattle Water Department's forecasting activities is the importance of carefully documenting forecast efforts. Documentation of the data bases used, techniques applied, and statistical results obtained—although time-consuming—establishes a necessary basis for future forecasts. Such documentation becomes critical when a forecaster needs to assess the applicability of various time-series analysis techniques to data series of changing length or needs to sort out record-keeping inconsistencies. For example, suppose that poor results are obtained with one technique and one data base, but better results are obtained in a later study (perhaps by a different analyst) with a different technique and a different data base. In such a situation, it is difficult to determine whether the improvement was due to the new technique or to the updated data base, unless the earlier technique can be applied to the new data base in exactly the same manner as before.

The further into the future a forecast extends, the more susceptible it becomes to errors in anticipating significant shifts in consumption patterns, income growth patterns, technological advances, and changes in relative prices. The increased potential for such errors reduces the likely long-term accuracy of the forecast. To deal with this uncertainty, several long-term scenarios can be analyzed in which key parameters are varied significantly. In a complex model with numerous parameters, distinct sets of assumptions may lead to the same forecast; perturbations in one variable may be offset by perturbations in another variable. Thus, multiple scenarios for long-term forecasts tend to exhibit diminishing returns to the forecasting process. Furthermore, the longer the time horizon, the less costly plan revisions and adjustments are. The returns on resources spent to increase the accuracy of long-term forecasts are therefore not great, and simple forecasting methods that incorporate a range of per-capita use factors and a range of population estimates are often adequate for the long term.

For other decisions in which actions with a time horizon of five to twenty years are contemplated, accuracy becomes more critical. The amount of time available for revising plans is reduced, and the costs of being wrong (for example, having to increase the size of a distribution main five years after it was installed, rather than installing a larger pipe initially) are closer in time and larger in terms of present value. In this situation, the return on resources spent for more comprehensive forecasting is probably greater than the cost of the considerable data-collecting efforts needed for such forecasts.

## REFERENCES

Baumann, D. D. 1981. Planning for water reuse. In *Selected Works in Water Supply, Water Conservation and Water Quality Planning*, ed. J. E. Crews and J. Tang, pp. 271–87. U.S. Army Corps of Engineers, Engineer Institute for Water Resources.

Boland, J. J., and Malloy, C. W. 1973. Comments on "Residential water demand forecasting" by Peter W. Whitford. *Water Resources Research* 9(3): 768–70.

Boland, J. J.; Moy, W. S.; Pacey, J. L.; and Steiner, R. C. 1983. *Forecasting Municipal and Industrial Water Use: A Handbook of Methods*. U.S. Army Corps of Engineers, Engineer Institute for Water Resources.

Bovet, E. D. 1973. *Evaluation of Quality Parameters in Water Resources Planning—a State-of-the-Art Survey of the Economics of Water Quality*. U.S. Army Corps of Engineers, Engineer Institute for Water Resources.

Collins, M. A., and Plummer, A. H., Jr. 1974. Industrial applications of Whitford's demand forecasting procedure. *Water Resources Research* 10(2): 345–47.

Crews, J. E., and Mugler, M. W. 1982. A national perspective on water conservation effectiveness. Paper presented at the AWRA annual meeting, San Francisco, California, October 1982.

Danielson, L. E. 1979. An analysis of residential demand for water using micro time-series data. *Water Resources Research* 15(4): 763–67.

DeKay, F. C. 1985. The evolution of water demand forecasting. *Journal of the American Water Works Association* 77(10): 54–61.

Dowdy, S., and Weardon, S. 1983. *Statistics for Research*. New York: John Wiley.

Dworkin, D. M. 1981. Municipal reuse of wastewater. In *Selected Works in Water Supply, Water Conservation and Water Quality Planning*, ed. J. E. Crews and J. Tang, pp. 289–95. U.S. Army Corps of Engineers, Engineer Institute for Water Resources.

Ernst and Ernst. 1973. *A Study of How Water Quality Factors Can Be Incorporated into Water Supply Analysis*. U.S. Army Corps of Engineers, Engineer Institute for Water Resources.

George, S. S. 1985. Energy forecasting techniques: an overview. In *Energy Forecasting: Proceedings of the Energy Division Session of the ASCE Conference in Detroit, Michigan, October 24, 1985*, ed. T. H. Morlan, pp. 12–30. New York: ASCE.

HUD. 1984. Residential water conservation projects—summary report. U.S. Department of Housing and Urban Development, Office of Policy Development and Research.

J. B. Gilbert & Associates. 1979. Water conservation for Nevada. Report prepared for the Nevada State Department of Conservation and Natural Resources, Division of Water Planning.

Jones, C. V.; Boland, J. J.; Crews, J. E.; DeKay, C. F.; and Morris, J. R. 1984. *Municipal Water Demand: Statistical and Management Issues*. Studies in Water Policy and Management, No. 4. Boulder: Westview Press.

Journal AWWA. 1985. Los Angeles residents paid to save water. *Journal of the American Water Works Association*, October 1985.

Kim, J. R., and McCuen, R. H. 1980. The impact of demand modification. In *Water Conservation Strategies*, pp. 53–56. Denver: AWWA.

Koenig, L. 1966. *Studies Relating to Market Projections for Advanced Waste Treatment*. Washington, D.C.: U.S. Department of Interior Publication WP-20-AWTR-17.

Landscape West and Irrigation News. 1980. Computerized irrigation. *Landscape West and Irrigation News* 3(9).

———. 1982. Renovation brings golf course back up to par. *Landscape West and Irrigation News* 5(9).

Mitre Corporation. 1972. *Water quality indices*.

Orange County Municipal Water District. 1985. Urban Water Management Plan (draft), July 1985.

Saleba, G. S. 1985. Water demand forecasting. Paper presented at the AWWA Seminar on Demand Forecasting and Financial Risk Assessment, Washington, D.C., June 23, 1985.

Whitford, P. W. 1973. Residential water demand forecasting. *Water Resources Research* 8(4): 829–39.

Yamauchi, H., and Huang, W. 1980. Organization and statistical analysis of water consumption data at the local level. In *Energy and Water Use Forecasting*, pp. 55–58. Denver: AWWA.

# CHAPTER THREE

# Water Pricing

Using advanced forecasting methods and techniques often raises issues of water pricing—in particular, the price elasticity of water demand. Of all the factors that affect water use, price frequently is the only one that the utility has the power to change; thus, it constitutes the only decision variable in many cases. Changes in water-rate level or design alter the prices users face at the margin, and thereby alter the level and pattern of water use (fig. 3-1). Understanding the interactions of these variables is essential to effective rate-making policy, as well as to supply planning.

This chapter presents a brief history of water-pricing, a description of current pricing policies, and a procedure for implementing a self-sustaining pricing policy (full-cost water-pricing). Next, the different rate and price structures used by water utilities are discussed; and finally, the impact of the price elasticity of water demand on the forecasting process is discussed, from the viewpoints of the economist and of the water planner.

## PRICING POLICIES

### HISTORY

The cost of providing water varies widely from one municipal water utility to another, even among utilities that have the same source of supply. Factors that affect this cost include the age of the system, the type of treatment needed, the costs of pumping, the size of the service area, and average and peak water demands.

The evolution of water pricing is shaped by these factors and by the physical and sociological environment surrounding the water utility. The price evolution traced by Goldstein (1986) for public water supply

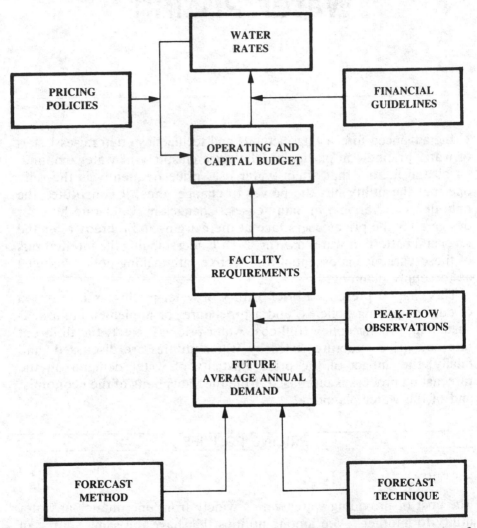

Figure 3-1. Flow chart of water-demand forecasting, showing the role of pricing policies in the forecasting process. (Adapted from figure on p. 31 of course outline for AWWA seminar on forecasting techniques, San Antonio, Texas, May 21–22, 1986; reproduced by permission, copyright © 1986, American Water Works Association)

systems in Massachusetts probably reflects the experiences of many water utilities established in times of water-supply abundance and cheap financing.

In the Goldstein example, the public water-supply system established in an area is designed to have substantial excess capacity, to take advantage of economies-of-scale and inexpensive water supplies in anticipation of rapid population and industrial growth. The early charges for water are deliberately set below the actual cost of supply in order to encourage water use, attract industry, ensure public health, and improve fire-fighting capabilities. By expanding the use of the water-supply system, the utility lowers the average cost (per unit) of supply to all customers. The utility subsidizes the cost of supplying water by drawing on general tax revenues; however, water-budget allocations are insufficient to cover the costs of necessary maintenance and capital improvement programs, resulting in substantial deterioration of the water-supply system. Unlike other municipal services, a poorly maintained water-supply or sewer system can go unnoticed by the public for decades.

During the 1970s, the cost of supplying water rises dramatically because of increased energy costs for pumping, increased quantity and price of chemicals for treatment, high inflation, and increased service costs for maintaining the aging infrastructure. The pressure on the already-strained water supply system is intensified by population growth and industrial development, as well as by the loss of supplies through contamination from hazardous wastes and other sources. The increased cost of providing water is only partly met by water-rate increases; the remainder is met by means of larger subsidies from tax revenues. Because of its aging system and inadequate revenues, the water utility defers necessary maintenance; as a consequence, fixed assets—especially water mains—continue to deteriorate.

In the 1980s, strict limitations are placed on property taxes, traditionally the major source of revenue for municipalities. Cutbacks in federal programs further reduce the funds available to cities and towns. In this environment, the utility begins considering substantial increases in water rates, with the following goals in view: to make its water system more nearly self-supporting and to alleviate pressure on limited general revenues; to provide revenues for long-needed maintenance and capital improvements; and to improve overall water management and system efficiency. This is precisely the situation many water utilities are facing now or may face in the future.

## CURRENT POLICIES

Goldstein cites five different types of institutional structures that may be responsible for setting water rates:

1. *City water department.* An elected city council appoints a city manager, who then appoints a water superintendent to operate the system. The city council sets water policy and water rates.

2. *Board of public works.* An elected board of selectmen appoints a board of public works, which in turn appoints a water superintendent to operate the system. The board of public works sets water policy, but water rates must be approved by the board of selectmen.

3. *Water and sewer commissions.* Water and sewer commissions are autonomous agencies established through special legislation, and these agencies take over responsibility for managing water supply from the municipality. Such commissions have been established in a few communities—mainly in response to financial pressures on municipal systems that could not cover the full cost of supply from existing revenues. Once a commission is established, it must be self-supporting, and this entails setting rates that cover the full cost of service.

4. *Water and fire districts.* Water and fire districts are established through an act of the state legislature and, by definition, are self-supporting. As autonomous bodies, they set rates based on the full cost of supplying water, and they retain all revenues collected. Such districts are usually found in suburban or rural areas, and a community may be divided into more than one district. As a result of their institutional structure and because most are less than fifty years old, water districts generally have less deferred maintenance than do municipal systems.

5. *Investor-owned water companies.* Although some private companies serve areas with populations greater than 10,000, most such companies are small and serve a few hundred customers or less. In order to remain in business, they must cover all costs and maintain a profit margin through their water rates. Unlike other water suppliers, private companies are monitored by regulatory agencies.

Regardless of the institutional structure used to set water rates, many water utilities face formidable obstacles when trying to secure adequate revenues to maintain and replace aging infrastructures. As described by Goldstein (1986), these obstacles are a lack of metering, historical underpricing, limitations of existing accounting systems, political concerns, and institutional constraints.

## Lack of Metering

Without meters, a utility cannot charge its customers equitably based on consumption. Some utilities meter most (but not all) users, with municipal departments remaining unmetered. Since municipalities are often large-volume water consumers, this can result in disproportionately high charges to other consumers or in increased subsidies from general revenues. Lack of metering also makes detecting leaks and tracing unaccounted-for water more difficult.

## Historical Underpricing

Consumers view water supply as a public service, similar to fire and police protection; consequently, they are unaware of the full cost a municipality must bear to supply water. In some cases, water service may be so underpriced that recovering the full cost of service requires steep (and unpopular) price increases.

## Accounting Systems

Budgeting and financial planning are often hindered by a lack of data on costs and revenues. For example, water-department capital expenditures commonly appear as an undifferentiated part of general-revenue bond issues that are included in debt-service accounts and not attributed to water-department costs. Without separate accounts for the financial transactions of the water department, it is difficult to calculate the cost of service and the degree to which revenues cover costs.

## Political Concerns

The primary sources of political pressure are individual consumers and industry. If no observable problem exists with the water-supply system, elected officials responsible for rate-making may be reluctant to support higher rates, regardless of the need. Numerous water-dependent companies, such as textile or paper mills, claim that higher water rates would force reductions in employment or even plant closures—neither of which is politically acceptable to elected officials.

Some experts believe that only a crisis, such as drought, a water ban, or a noticeable deterioration in water quality, will clear the way for public acceptance of increased water rates.

### Institutional Structure

Since many municipalities already lack adequate funds to operate and maintain their water systems properly, local water utility managers are understandably reluctant to spend a significant amount of their limited funds on administrative changes. If current practices are perceived as having effectively met past needs, they often become entrenched and difficult to change.

## SUBSIDIZED VERSUS SELF-SUSTAINING PRICING POLICIES

The pricing policies currently in use bear evidence that, historically, water rates have not been based on the full cost of providing water (Goldstein 1986). Instead, a water utility's funding typically is based on municipal budget allocations unrelated to the goal of recovering the full cost of service. In some instances, communities use water revenues to subsidize the cost of providing other municipal services; but more often a utility contributes less to the general fund than it receives from the budget-allocation process, and thus is subsidized by tax revenues. Figure 3-2 shows a schematic representation of a subsidized water utility.

The result of such practices is that rate levels do not meet the full cost of supplying water, yielding inadequate revenues for system maintenance and capital renewals, inequitable rate structures, and insufficient incentives for conservation. Although short-term savings may be realized through a policy of not investing continually in the supply system, long-term costs grow substantially as a result of the postponement of improvements, the accelerated deterioration of fixed assets, and the lack of overall planning.

In a full-cost pricing system, price levels are set to recover all of the direct and indirect costs associated with providing water, including capital expenditures, depreciation, billing and administration costs, and costs of services provided to the water system by other municipal departments (fig. 3-3).

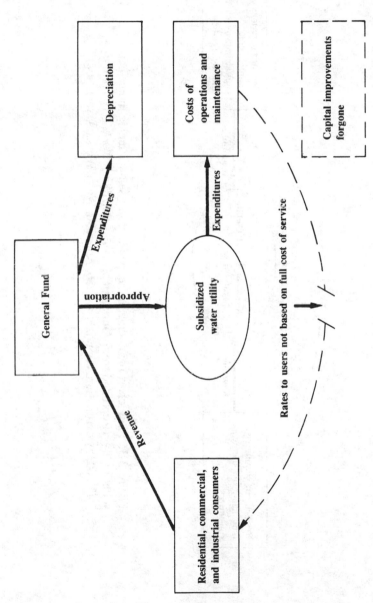

Figure 3-2. Flow chart of funding and expenses for a subsidized municipal water utility. (Reprinted from *Journal of the American Water Works Association* 78(2): 55, by permission; copyright © 1986, American Water Works Association)

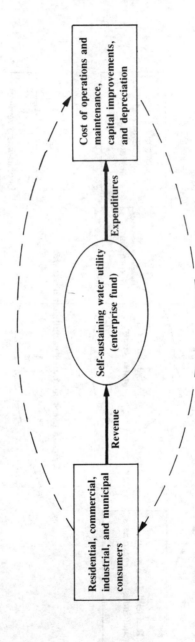

Figure 3-3. Flow chart of funding and expenses for a self-sustaining municipal water utility. (Reprinted from *Journal of the American Water Works Association* 78(2): 60, by permission; copyright © 1986, American Water Works Association)

The following benefits are possible from full-cost pricing:

**1.** Funds are available for regular maintenance, thereby reducing water loss and prolonging the life of the system's fixed assets.
**2.** Wasteful use is discouraged, and savings in water use may postpone, reduce, or eliminate the need for a new source of supply or a wastewater-treatment facility.
**3.** The pricing system is more equitable if all users are charged on the basis of the actual quantities of water they use.
**4.** Sufficient funds are available to protect sources and watersheds.
**5.** Incentives for sound financial practices and efficient management are introduced.
**6.** Sufficient continuing revenues are provided to ensure a reliable, high-quality supply.
**7.** Consumers develop an awareness of the true monetary value of water supply.

## PROCEDURE FOR IMPLEMENTING FULL-COST PRICING

For a utility to become self-sustaining and to recover the full cost of providing water, Goldstein recommends that it take the following seven steps.

**Step 1: Establish Enterprise-Fund Accounting.** The practice of keeping a municipal utility's costs and revenues separate from the general account is referred to as enterprise-fund accounting. If water utility costs and revenues are lumped together with other municipal expenditures and receipts, accounting records do not keep track of when water revenues have been diverted to pay for other municipal services or when they have been inadequate to cover all water-utility costs. Enterprise-fund accounting provides the financial information necessary to determine the full cost of water service and the extent to which revenues generated from rates cover this cost.

**Step 2: Calculate the Full Cost of Service.** The full cost of providing water includes such expenses as capital improvements and depreciation of fixed assets. Calculating this cost identifies the revenue needs of the water utility. A general list of the costs borne by a municipal water utility might include the following items: land acquisition, management, and protection; fixed assets, including reservoirs, wells, pumping stations, water mains, treatment plants, and storage tanks; water itself, if purchased from a wholesaler or another supplier; maintenance of fixed assets, including equipment and materials; pumping costs (energy); labor, including employee fringe benefits, overtime, and emergency and part-time personnel; treatment costs (chemicals); planning and supply augmentation, including contracts for emergency supplies and consultants; office overhead, including electricity, heating, equipment, and supplies; insurance for fixed assets, vehicles, and equipment; billing and administration, including computer services, meter-reading, and services provided by other municipal departments; fire protection services; and depreciation.

**Step 3: Evaluate Fixed Assets and Determine Depreciation.** The extent to which revenues generated from water rates cover the full cost of service should be determined. Ideally, an exact valuation of assets should be performed, to provide the basis for calculating depreciation costs. The valuation of assets also uncovers useful data on the present condition and future needs of the water-supply system.

**Step 4: Increase the Use of Meters.** The incoming water-supply lines of all consumers, including municipal users, should be metered. A municipal water supplier can charge equitable rates for water use only if all of its customers are metered. In addition, 100-percent metering is necessary for tracing unaccounted-for water and for detecting leaks, thereby contributing to water conservation.

**Step 5: Educate the Public.** An education campaign should be undertaken, to inform the public and local officials about the full cost of providing water and about the financial management practices required for meeting this cost. The public-education effort should not be a general campaign on the virtues of full-cost pricing; instead, it should identify precise costs and system needs, and it should link these to specific policy proposals.

**Step 6: Improve the Rate-making Process.** The rate-making process can be improved by minimizing political considerations and focusing instead on two areas: increased efficiency in operations and maintenance, and the capital improvement needs of the water-distribution system. This might involve appointing a special board or committee to examine needs, recommend appropriate water rates, and formulate related policies.

**Step 7: Achieve Full-cost Pricing.**   A water utility wanting to implement full-cost pricing may do so either through one price increase or through several incremental increases. Most water utilities plan their rates to cover a two- to four-year time span, although some adjust rates annually. Public opposition and reduced consumer confidence in the utility may result from frequent price hikes. In such cases, it is advantageous for the utility to raise its prices only once within a two- to four-year period. On the other hand, achieving full-cost pricing with one large price increase may be unacceptable to local officials and consumers. In such cases, the utility can institute incremental price increases, making clear at the outset that its goal is to be able to recover the full cost of service from rates by a specific date and that a series of increases is planned.

## WATER RATES AND PRICE STRUCTURES

Goldstein (1986) identifies two components of water rates: the price structure, and the price level. The *price structure* is the method of charging for water; it is based either on the quantity used or the time of use (fig. 3-4). The *price level* is the actual price charged for a given volume of water. Although price structure and price level are in theory mutually independent, they are often considered together in the rate-setting process. Any water rate can recover the total cost of service, as long as the price level is high enough.

Water utilities often include a minimum charge in their water rates, regardless of the price structure used. The minimum charge covers per-period use by the consumer of quantities up to a specified volume of water; water use greater than this volume is charged according to the price structure in effect. A minimum charge is useful because it ensures a certain level of income from all customers. It is to some extent inequitable, however, since consumers who use less than the volume allotted must still pay the minimum charge. Minimum charges are increasingly being replaced by service charges. Service charges cover all of the administrative cost of maintaining water service, but—unlike minimum charges—they do not penalize low-volume consumers by charging them for unused water.

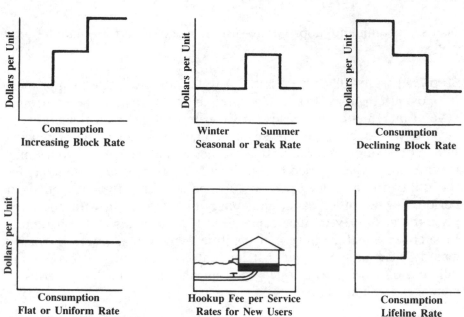

Figure 3-4. Common pricing structures for water services. (Reprinted from *Journal of the American Water Works Association*, 78(2): 58, by permission; copyright © 1986, American Water Works Association)

Retail water agencies can use a number of water-pricing structures, including fixed charges, uniform rates, varying rates, peak-load pricing, and marginal-cost pricing. The advantages and disadvantages of these structures are discussed in the following sections (Goldstein 1986; Craddock 1985; Orange County Municipal Water District 1985).

## FIXED CHARGES

Although most communities with municipally supplied water are metered, some communities (usually small towns) are not. The most common method of charging for water service in municipalities that have little or no metering is to set a uniform charge for each connection or to establish a fixture rate, whereby charges are based on the number of water fixtures in the household. In unmetered water service, no relationship exists between consumption and the charge for water; as a result, communities lacking metered service cannot use pricing to encourage conservation or to achieve other policy goals.

## FLAT OR UNIFORM RATE

A flat or uniform rate is simply a constant price per unit of volume for all water use by all consumers. The flat rate is easy to design and administer and is considered by many to be the most equitable method of charging because all consumers are charged the same rate per unit. Large-volume users, particularly industrial users, argue that the flat rate is inequitable to them because they tend to equalize their water use over the course of the day or year and so use a smaller proportion of their water during times of peak demand. Since the investments in additional supplies and storage capacity needed to meet peak demand add substantially to the cost of service, large-volume users argue that they are less responsible for these costs and should not be required to pay for them.

## VARYING RATES

In water-pricing, the additional unit representing a specific quantity of water is called a *block*. Varying rates can follow either a declining-block schedule or an increasing-block schedule.

### Declining-Block Rate

Until recently, the most typical rate structure for water service was the declining-block rate. Under this system, customers usually pay a minimum charge and are entitled to use a specific amount of water without additional cost. Water use greater than this specified amount is divided into blocks, each of which is priced at a lower rate. This form of pricing dates back to a time when most water systems had substantial excess capacity.

The traditional justification for implementing declining-block rates is that water use by small-volume customers is more concentrated during peak periods than is water use by large-volume customers. A large part of the cost of delivering water is taken up by the cost of providing sufficient capacity to handle the peak load on the distribution system. Consequently, the needs of large-volume customers are less costly to supply because they involve less of a contribution than the needs of small-volume customers do to the peak load on the system. Furthermore, low-cost water provides an economic incentive to business users, with the community ultimately benefiting from a healthy local economy.

Critics argue that this system encourages excessive consumption. As the price per block decreases, the incentive to conserve decreases. Furthermore, large-volume residential customers, who use large amounts of water for outdoor landscaping, contribute more to the peak load than do small-volume residential customers. The justification for giving quantity discounts to large-volume users falls apart when applied to them.

### Increasing-Block Rate

The structure of increasing-block rates provides that each succeeding block of water used is priced at a higher rate per unit than the previous block. As in other rate structures, a minimum charge entitles customers to a specific volume of water—the first block. This type of rate is often used if the utility is faced with a supply shortage or if water conservation is a critical concern. The increasing-block rate structure can be designed to allow for essential domestic use at lower rates (similar to lifeline rates for electricity), with higher rates for large-volume consumers.

Conservation is encouraged in this pricing system. The lowest per-block cost applies only to the first block consumed during the billing period; progressively higher fees are assessed for subsequent blocks. One objective of this pricing system is to reduce summer sprinkling demand among residential users. In this application, the first block quantity approximates winter levels of use. The incremental cost of subsequent blocks may motivate consumers to install water-saving devices or drought-resistant landscaping and to practice conservation year-round. Under normal weather conditions, revenue would cover the operating costs, and peak demand in summer would be reduced, rendering the system's capacity utilization more efficient year-round.

An increasing-block rate schedule can cause large-volume industrial consumers to pay a high price for water, even though their water use is not concentrated in the peak periods. This problem can be avoided by having a separate rate schedule for industrial consumers, or by having a humpback rate schedule in which the price of water falls again at some level that exceeds monthly use by most residential customers. In its nondifferentiated form, this rate structure may be best-suited for primarily residential areas or for communities trying to discourage industrial development. Increasing-block rates may be an unstable source of revenue, since reductions in water use occur at the highest rate block.

Although the increasing-block rate system is a desirable response to increased water costs and demand, implementation of the program requires extensive planning. An effective structure of block quantities, levels, and rates must be based on a thorough analysis of consumption patterns among users in each area and of the revenue requirements of water suppliers.

## PEAK-LOAD PRICING

Since the need to enlarge a distribution system results from the summer peak load on the system, summer use imposes greater costs on the utility than does winter use. In the long run, the costs from summer use comprise the cost of enlarging the system to provide for peak loads, plus the cost of operating the water system, plus the cost of the water itself. The costs from winter use consist merely of the cost of operating the water system and the cost of the water—because the distribution system has already been built to provide for the higher levels of summer use. Consequently, any rate schedule that charges the same price for water all year tends to underprice water used in the summer and overprice water used in the winter.

Peak-load pricing schedules either use a surcharge on water use during high-demand periods or use seasonal rates. When a surcharge schedule is adopted, uniform or varying rates are charged for a base amount of water, and higher rates are charged for amounts of water exceeding the base amount. The higher rates apply to water use during periods of peak demand and may be in force during the summer sprinkling season only or throughout the year. These rates reflect the utility's large investments in storage and other facilities to meet peak use, which is usually the determining factor in sizing system capacity.

Seasonal pricing uses a divided rate structure, with higher per-unit charges in summer months than in winter months. Compared to a surcharge, seasonal pricing is simpler to establish, easier to explain to consumers, and less complex to administer. The seasonal or peak-use rate can be used in conjunction with any price structure, but it is most commonly used with the flat rate. The peak-use rate can be applied to all water used during the peak period or only to the portion that exceeds a predetermined base amount such as average (off-peak) use. The latter method allows customers to meet their basic water needs without having to pay a penalty.

Russell (1984) noted the following four objectives underlying adoption of peak-load pricing:

**1.** *Cost recovery.* Higher rates during the summer season may reflect a more equitable means of recovering the cost of providing water service by singling out customers who use more water than average during the summer.

**2.** *Peak-demand reduction.* Higher prices in summer could reduce peak-day and peak-hour demands, thus postponing or eliminating the need for construction of additional capacity.

**3.** *Extending available supplies.* In situations where the supply is limited or the development of additional sources is too expensive, the seasonal rate may postpone or eliminate the need for a major expansion of the system.

**4.** *Conservation.* Higher prices in summer may encourage conservation and better utilization of the water supply, leading to savings in natural resources, energy, and chemicals.

One substantial problem with peak-load pricing involves rate stability. The revenue requirements for water systems are relatively fixed and do not vary with changes in volume. If a seasonal rate is in effect, however, significant increases or decreases in revenue can result from changes in sales volume caused by weather fluctuations. Thus, the timing of the implementation of peak-load pricing is important. If peak-load pricing is instituted in the fall, for example, the utility may have a revenue shortfall before the next summer arrives. Peak-load pricing should probably go into effect at the start of summer to ensure adequate cash flow. Administering peak rates may also require more frequent meter-reading.

The principal objection to peak-load pricing lies in defining the objective of water-pricing policy: is it in the best interest of the service area, and what are the impacts of such seasonal prices on the area? Seasonal variations in water rates could be likened to the situation that would exist if the Postal Service boosted the price of postage required to mail a letter during the busy Christmas season, in order to curtail the mailing of Christmas cards.

## MARGINAL-COST PRICING

Extending a water-supply system to new users is usually expensive. The extensions may be paid for in a variety of ways, depending on the policy objectives of the municipality. For example, a community wanting to encourage new residential or industrial development may sell bonds to pay for the project, and then recoup the cost from all system users by adjusting the rates so that charges for principal and interest are included in the cost-of-service calculation.

A community that wants to discourage new development may choose to charge the developer or builder for the total cost of extending the system. Requiring payment of part or all of this cost before building begins is often enough to dissuade potential developers. Developers who are willing to accept these terms usually pass this cost on to the future property owners.

A community unopposed to development that nonetheless does not want to charge present users for the cost of extending its water system may recoup these costs from the new users directly. Usually this involves imposing a hookup charge that may cost thousands of dollars per new user. This method of charging for system extension (called *marginal-cost pricing*) is considered the most equitable system by many people, since new users pay for the full costs of extending the system, and the rates for existing users are not affected.

Marginal cost—the additional cost of producing and selling a single additional unit of output—is a function of output or amount of water produced. When additional blocks can be supplied at lower per-unit costs than previous production, marginal cost is lower than average cost. If incremental units of water are more costly to produce than preceding units, however, marginal cost will exceed average cost. Theoretically, marginal-cost pricing increases efficiency in water usage, since higher prices for incremental units should induce many consumers to conserve. Under these circumstances, only customers who place a high value on water are expected to use large quantities of water. The marginal-cost principle also implies that, if service to a user involves extra cost, the user should pay those added costs and not be subsidized by other water users.

Marginal-cost pricing raises a number of problems. Because they are future costs, marginal costs require additional forecasts of costs and demand. Marginal-cost pricing causes fluctuations in price, capacity utilization, and revenue—uncertainties that are compounded by the effect of weather on water demand. Changes in income distribution may occur if water usage is shifted from the poor to the rich or if an industry chooses to relocate.

## SUMMARY EVALUATION OF PRICING STRUCTURES

No pricing structure is ideal for all communities. The most appropriate price structure depends on local conditions and policy objectives—namely, what the municipality wants its water rates to accomplish. As shown in the previous examples, rates can be structured to encourage water conservation, reduce peak demand, attract industry, or help meet other goals. The effectiveness of using water rates and pricing structures as incentives for conservation is summarized in figure 3-5.

In evaluating the appropriateness of a water rate, Goldstein (1986) suggests planners should ask the following questions:

1. Does the rate provide adequate revenues to cover all costs of service?
2. Are costs fairly apportioned to each class of consumer? That is, is the rate equitable to residential, commercial, and industrial users?
3. Are the rates acceptable to the public and to local officials responsible for administering them?
4. Does the rate encourage or discourage water conservation?

Since water utilities usually have high fixed costs and relatively low variable costs, conservation of water—even when spurred by increased price levels—may adversely affect the water utility's cash flow. Depending on the extent of conservation that results from a rate increase, total water utility revenues may increase or decrease: if little conservation is stimulated by the rate increase, total revenues should increase; but if a significant rise in conservation occurs, total water utility revenues may decline. Legislation recently introduced in New York is designed to prevent water companies from raising rates to recoup their losses when water consumption is reduced through conservation. As one legislator has said, "It's like asking someone to help you lift a heavy load and then picking his pockets when his hands are busy" (*Engineering News Record* 1986). A water utility considering increasing its water rates should not proceed without a good estimate of the effect the increase will have on water use.

---

Figure 3-5. Conservation incentives provided by different water-rate schedules. (Reprinted from *Supplying the Demand: The Water Management Challenge*, 1984, by permission of the Freshwater Foundation)

# WATER RATES AS INCENTIVES TO CONSERVE

| Type of Rate Schedule | Incentive To Conserve Provided |
|---|---|
| 1. FIXED CHARGE: Any particular customer is charged the same dollar amount per period regardless of the quantity of water consumed. Some form of fixed charge must be used for customers without water meters. | Provides no incentive to conserve because the quantity of water used is not related to the price of water. |
| 2. UNIFORM RATE: The unit rate is the same for all units of water consumed. One *unit* is typically 1,000 gallons or 100 cubic feet (750 gallons). This is equivalent to a single price. | Fosters water conservation in a manner that does not vary with increased consumption. |
| 3. VARYING RATE: Can either be a declining block or an increasing block schedule.<br><br>*Declining block:* A certain rate per unit is charged for all units consumed up to a specified quantity. Such a bounded consumption range is known as a "block." For water consumed in the next block, a lower unit rate is charged. In this manner, unit rates *decrease* for succeeding consumption blocks.<br><br>*Increasing block:* This is the reverse of the declining block type, in that unit rates *increase* with each succeeding block. | Provides a conservation incentive that *diminishes* as consumption increases. |
| 4. PEAK LOAD PRICING: Uses either a seasonal rate or a surcharge.<br><br>*Seasonal rate:* Higher unit rates are imposed in the summer, when demand increases due to lawn sprinkling. The unit rates can be either uniform or varying.<br><br>*Surcharge:* Uniform or varying rates are charged for a base amount of water. Higher rates are charged for water in excess of the base. These higher rates can be applied either during the summer sprinkling season or throughout the year. | Have the most potential for inducing water conservation, by creating an incentive that *increases* with greater water use. This type of incentive is likely to be most effective in reducing the lawn-sprinkling component of water use. |
| 5. MIXED: Most rate schedules in effect today are mixed in the sense of combining unit rates with a flat charge. The flat rate can either be a minimum charge or a fixed charge. A minimum charge includes a certain allotment of water, beyond which unit rates apply. | Depends on the type of unit rate in the schedule, except when consumption lies within the minimum-charge block. In this case, no incentive is provided to conserve, since the homeowner must still pay the full minimum charge. |

## The Need for Redesigning

For a sample of 90 water rate schedules drawn from a survey by the American Water Works Association, over half the utilities (55.6 percent) use the minimum charge/declining block type of schedule, and another 25 percent use the minimum charge/uniform rate type. Furthermore, *less than 6 percent of the schedules can provide a conservation incentive that in-creases with more water use* (fixed/increasing, fixed/seasonal and minimum/increasing). These figures suggest that redesign of rate schedules could lead to increased water conservation.

Excerpted from "Cost-Effective Residential Water Conservation Decisions," presented at the National Water Conservation Conference on Publicly Supplied Potable Water, Denver, Colorado, April 1981, by Stephen F. Weber, Barbara C. Lippiatt and Anne P. Hillstrom.

## PRICE ELASTICITY OF DEMAND

In discussions of the relationship between price and water, a fundamental difference of opinion is often encountered between the water planner and the economist. The economist will point out that, while water is essential to human life, the quantity required to sustain life is small (less than 2 liters per person per day) and can easily be supplied by means other than public distribution systems (in food, as bottled water, in soft drinks, and so on). Therefore, economists claim that water is purchased and used in a way that does not differ fundamentally from the way bread, gasoline, or any other staple commodity is; consequently, water use can be affected by the price charged for it. In comparison, many water-utility managers consider that the price of water has little, if any, effect on the demand for water.

## THE ROLE OF PRICE

In chapter 1, major factors (explanatory variables) affecting water use were identified, and their relationships to each component of urban water use were expressed in quantitative terms. For example, explanatory variables of residential water use include (among others) number of households, population per household, household income, property value, irrigable area, and climate. Explanatory variables of industrial water use include such things as employment, industrial output, and recycle ratio; and explanatory variables of commercial water use include layout and design characteristics, transportation balance, and labor productivity.

There are two general categories of explanatory variables: variables that determine the need for water, and variables that determine the intensity of water use. *Need* variables include population served, number of households, and industrial employment. The presence of these factors indicates that water-using activities are occurring and that some water is required; the amount of water needed, however, is not clear from these factors.

*Intensity-of-use* variables include income (ability to pay for water), conservation practices (willingness to substitute inconvenience or other inputs for water use), and price (willingness to pay for water). For a given set of water-using activities, water use will grow with increasing income, diminish with increasing conservation activity, and diminish with increasing price.

In comparison to such variables as number of households, population per household, and climate, price accounts for relatively little variance in water use. Yet variations in price have been responsible for significant shifts in use levels. Unlike most other factors, price can either increase or decrease and can do so abruptly and substantially. Because of the nature of the relationship betwen price and use, adjustments to changes in price are not instantaneous. A change in price induces a slow and steady change in water use, that may be completed only after a period of up to ten years. These characteristics give price (as a forecasting parameter) an importance that goes beyond its basic explanatory power.

## DEFINITION OF PRICE ELASTICITY

Economists define *water demand* as the relationship between water use and price, when all other factors are held constant. Demand is a negative functional relationship, illustrated by the demand curve in figure 3-6 that plots the relationship between price and water use for a single user. The demand imposed by every water user can be represented by a series of similar demand curves, each of which will presumably have a negative slope (because increased price results in decreased water use).

When a group of users faces a price that is uniform over the group, the users' individual demand curves can be summed horizontally to obtain an aggregate demand curve, as shown in figure 3-7. The aggregate demand curve, usually called a *market-demand curve* also has a negative slope. Clearly, there exists a price $P'$ at which no one will purchase water from the public system (because everyone will prefer to obtain water by other means). And, in the event that no price is set (so that price equals zero), there nonetheless exists some finite maximum quantity of water, $Q'$, that will be demanded. Between these two extremes, the quantity of water demanded is determined by the price and by the demand curve, if all other factors are held constant.

The shape and position of the demand curve are determined by the values of the other explanatory variables (including the need variables), as well as by income and conservation practices. The effect of increasing income is to shift the curve to the right (fig. 3-8), so that—at the

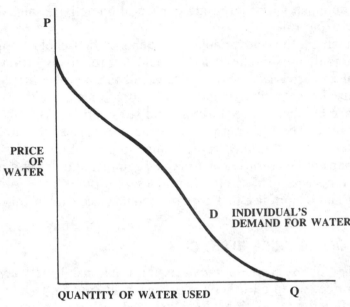

Figure 3-6. Individual demand curve. (Reprinted from U.S. Army Corps of Engineers 1984)

same price, $P_0$—progressively larger quantities of water are used ($Q_1$, $Q_2$, $Q_3$). The effect of increasing conservation is to shift the curve to the left (fig. 3-9). Increasing the levels of the need variables normally moves the demand curve to the right. All of these shifts may be accompanied by changes in the shape and slope of the demand curve.

Water-supply planning rarely requires that the entire demand curve be known. More often, knowing how specific incremental changes in explanatory variables will affect water use suffices for planning. In the case of price, this information is represented in the slope of the demand curve, which gives the incremental change in use for an incremental change in price at some position on the curve (fig. 3-10).

Because the units chosen for the axes of the demand curve are dollars per unit of water use, and units of water use, the slope of the curve has an inconvenient dimension (dollars per unit of water use squared). It is customary, therefore, to use a dimensionless measure of the relationship, obtained by dividing fractional (instead of incremental) change in water use by fractional change in price. This dimensionless measure is known as an *elasticity*. The *price elasticity*

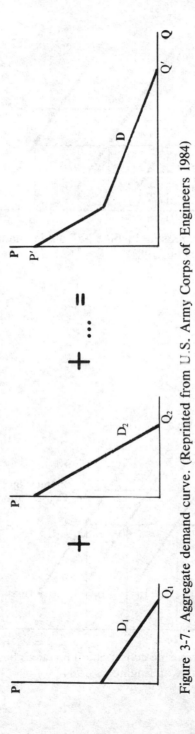

Figure 3-7. Aggregate demand curve. (Reprinted from U.S. Army Corps of Engineers 1984)

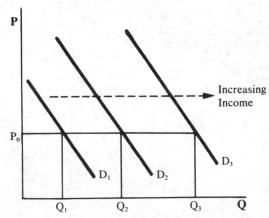

Figure 3-8. Effect increasing income has on demand. (Reprinted from U.S. Army Corps of Engineers 1984)

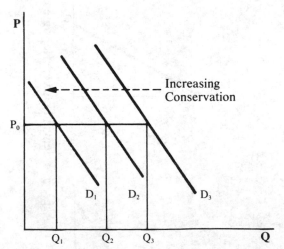

Figure 3-9. Effect increasing conservation practices has on demand. (Reprinted from U.S. Army Corps of Engineers 1984)

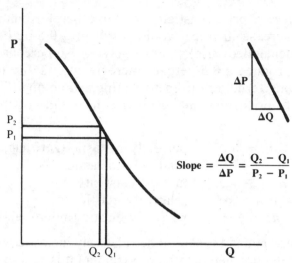

Figure 3-10. Defining the slope of a demand curve. (Reprinted from U.S. Army Corps of Engineers 1984)

*of water demand* is defined for an arc of the curve, as shown in figure 3-7, as:

$$n = [(Q_2 - Q_1)/Q^*]/[(P_2 - P_1)/P^*] \qquad (3.1)$$

where:

$$Q^* = (Q_2 + Q_1)/2$$
$$P^* = (P_2 + P_1)/2$$

As shown in equation 3.1, price elasticity of demand is expressed as the ratio of percentage change in quantity demanded to percentage change in price. The use of percentages makes price elasticity independent of the units used to define price and quantity. For example, it makes no difference if the price of water is expressed in cents per 100 cubic feet or in dollars per acre-foot, or if the quantity demanded is expressed in gallons per capita per day or in acre-feet per service connection per year.

Elasticity is expected to be a negative quantity because the demand curve has a negative slope. Price elasticity may be interpreted as the percentage change in demand quantity that would result from a percent change in price. A price elasticity of $-0.5$ therefore indicates that a 1.0-percent increase in price would result in a 0.5-percent decrease in quantity demanded (use). Conversely, a 1.0-percent decrease in price would produce a 0.5-percent increase in quantity demanded.

The following terms are used to distinguish among different types of response to price, depending on the magnitude of the calculated elasticity:

| | |
|---|---|
| $n = 0.0$ | perfectly inelastic (zero elasticity) |
| $0.0 > n > -1.0$ | relatively inelastic |
| $n = -1.0$ | unitary elasticity |
| $-1.0 > n > -\infty$ | relatively elastic |
| $n = -\infty$ | perfectly elastic (infinite elasticity) |

In other words, demand is said to be relatively inelastic when quantity changes disproportionately less than price, and it is said to be relatively elastic when quantity changes disproportionately more than price (fig. 3-11).

If the demand for a commodity is inelastic, an increase in the price of the commodity will result in an increase in total revenues. This is because the percentage increase in price is greater than the percentage decrease in demand for the commodity. On the other hand, if the demand is elastic, an increase in price will result in a decrease in total revenues because the decrease in demand exceeds the increase in price.

For example, suppose that a commodity has a price elasticity of 0.5 and initial sales of 1,000 units at $10 per unit. If the price is increased to $12 per unit (a 20-percent increase in price), sales will decrease by 10 percent to 900 units. The total revenues, which were originally $10,000 (1,000 units × $10 per unit), will increase to $10,800 (900 units × $12 per unit).

Since price elasticity of demand is defined at a particular point along a demand curve, a different price elasticity value may be found at another point. If the demand curve is linear, for example, price elasticity values grow increasingly negative at progressively higher prices, and vice versa (fig. 3-12). Equally plausible demand curves can be constructed that maintain the same elasticity at every point (fig. 3-13) or have an elasticity that grows increasingly negative at progressively lower prices. Elasticity may also increase, decrease, or remain the same in response to decreasing price levels.

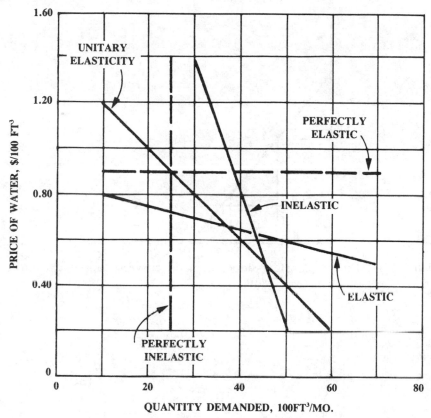

Figure 3-11. Elasticity of linear demand curves. (Adapted from Hildebrand 1984)

## FACTORS INFLUENCING PRICE ELASTICITY

Many factors can influence price elasticity. A literature survey conducted by Boland et al. (1984) for the U.S. Army Corps of Engineers cites the following factors that could influence price elasticity: user class, season, changes in explanatory variables, user response, and rate-structure characteristics.

### User Class

Systematic differences in price response can be observed among user classes. When price response is measured at too high a level of aggregation, these differences become submerged in the data, and the

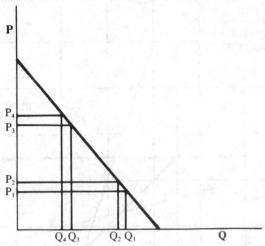

Figure 3-12. Linear (constant-slope) demand curve. (Reprinted from U.S. Army Corps of Engineers 1984)

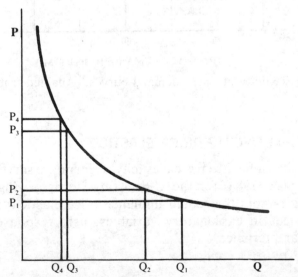

Figure 3-13. Exponential (constant-elasticity) demand curve. (Reprinted from U.S. Army Corps of Engineers 1984)

result is an elasticity that at best represents a weighted average of the component elasticities.

The best example of an aggregation problem is the practice of analyzing average annual urban water use. The resulting elasticity value is a weighted average of residential, commercial, institutional, industrial, and other elasticities, as well as of summer and winter elasticities for each of the classes. Since the weights vary from community to community (because of different proportions of users in each class and different weather patterns), the aggregate elasticity varies, too. Such results may be useful in the community for which they are derived, but they are not usually transferrable to other communities.

The most widely applicable estimates of price elasticity, therefore, are ones that apply to the smallest and most homogeneous classes of water use. In the case of residential use, these would include estimates of winter (or nonseasonal) and summer (or seasonal) elasticities. In the case of industrial or commercial water use, estimates of elasticity for specific categories (poultry processing, department stores, and so on) are preferable to estimates for the class as a whole. Using higher levels of aggregation introduces study area–specific variations into the estimates, producing a broader range of results but making application to other areas more difficult.

## Season
Even though user classes are intentionally defined to be as homogeneous as possible, many different uses—affected by different explanatory variables—still occur within each class. One method of further clarifying basic relationships is to perform separate analyses of summer and winter (or seasonal and nonseasonal) water use within a class. Doing this isolates the relatively homogeneous winter (or nonseasonal) water use from the summer (seasonal) use, which includes various irrigation and other outside uses.

Since the components of water use vary by season, the relevant explanatory variables and their relationships to water use vary as well. Price elasticity of demand, therefore, can be expected to vary between summer and winter (or seasonal and nonseasonal) uses. Since studies have shown relatively large differences between elasticities for residential winter use and those for residential summer use, the practice of analyzing average residential use without regard to season conceals the true components of price response.

## Changes in Explanatory Variables

One of the most important sources of bias in elasticity studies (besides sample selection and data measurement) is the choice of which explanatory variables to consider in the regression model. Price is commonly collinear with other variables, and the omission of those variables may produce a biased coefficient.

For example, price is usually lower in larger communities, which also contain relatively large numbers of multiunit residential buildings. If per-capita water use is taken as the dependent variable, it could be expected to go up if a larger fraction of the population lived in smaller household units or if the price were lower. Omission of explanatory variables describing household size and fraction of multiunit housing would result in both effects' being reflected in the price coefficient, leading to an overestimation of the price elasticity.

## User Response

Elasticity also varies according to the users' ability to make cost-effective adjustments in utilizing related goods and in water-using habits. In the case of water, adjustments may include changes in the stock of water-using appliances, changes in landscaping, changes in irrigation practices, and changes in domestic water-use habits. When the user is free to adjust any related good or behavior, the measured adjustment to price is described as a long-run elasticity. When one or more of the adjustments is not available for some reason, the adjustment is measured as a short-run elasticity.

Adjustments to water price all require the passage of time—perhaps up to a decade to complete changes in the stock of major water-using appliances, for example. Long-run responses are likely to be more elastic than are short-run responses, although the difference may not be large in every case. A number of years may have to elapse before the long-run response can be presumed to be complete. A short-run response may be fully instituted within weeks or months of the effective date of a price change.

Exceptions to these generalities should be noted. Changes in water price do not affect all users simultaneously. Typically, the utility announces that a change will be put into effect for all meter readings or bills occurring after a certain date. Depending on the meter-reading cycle and the billing lag, it may be four months or more before all customers actually receive a bill calculated according to the new rates.

Two different (and sometimes contradictory) responses may be observed. First, some users may react immediately upon hearing of the new rate, even before it actually takes effect. This response—the announcement effect—is based on the expected, rather than actual, impact of the new rates. To the extent that the change's perceived impact is greater than its actual impact, the initial response may be greater than the later net adjustment. On the other hand, if the initial expectation underestimates the impact of the rates, the first response may represent only a portion of the entire adjustment ultimately adopted.

Second, other users may ignore or be unaware of the announcement, postponing their response until the first bill is received at the new rates. On seeing the impact of the rates, they may undertake a series of short-run, and then long-run, adjustments as described above. Prior to receiving the first bill, however, users in this group have made no change in their water-use patterns, even though the new price is in effect.

Attempts to observe short-run elasticities by means of time-series analyses that extend over periods of less than one year may be confounded by these complicating factors. Some users may have reacted to the announcement by overestimating or underestimating the actual impact of the price change; others may not react at all until the first bill arrives, at which point their reactions begin gradually to be phased into overall water-use statistics as the meters are read and the bills rendered. The observed progression from an initial short-run to a long-run response may itself be distorted by the billing cycle. While not critically important for long-range forecasting or demand modeling, short-run elasticity estimates are very useful in rate design and revenue-forecasting. Short-run estimates may also be relevant to drought-management planning, where the short-term response to emergency price changes is of interest.

## Rate Structure Characteristics

While most economic goods are sold to consumers at well-defined prices, water is priced according to relatively complex rate schedules. These schedules may include a number of fixed charges (including assessments, service charges, and minimum charges), as well as variable charges. The variable charges may differ from one use to another (decreasing- and increasing-block rates) or from one season to another (seasonal rates).

Economic theory states that the price that affects the level of use is the price paid at the margin—that is, the price paid for the last unit used. Depending on the structure of rates, this price may vary from user to user or from time to time for the same user. It may be difficult or impossible to determine the marginal price associated with each observation of water use. For example, when water use is aggregated over a number of users who face block-type rates, marginal-price data are inevitably lost. For these reasons, many studies rely on measures of average price, sometimes calculated by dividing total revenue from charges by total water sold.

When time-series data are used, price data must be deflated to a constant dollar measure by means of a suitable index. National or local consumer price indices are most often used for this purpose. In the case of seasonal rates, the forecaster may have to develop measures of price that account for lags in the billing cycle and for the perceptions of users regarding cyclical changes in price.

Correctly specifying price is of fundamental importance in estimating price elasticities. Even when price has been correctly specified, however, the characteristics of the rate structure may introduce bias. For example, when decreasing-block designs are used, the marginal price decreases as more water is used. This ensures a negative functional relationship between price and use, even if customers are completely insensitive to price. Data collected from individual customers facing such a rate, therefore, inevitably overestimate price elasticity. Conversely, increasing-block rates produce underestimations of price elasticity.

## METHOD FOR DETERMINING PRICE ELASTICITY

Estimates of price elasticity are obtained through econometric analysis of price/quantity data for representative water users. In order to interpret the data, the existence and the specific functional form of a demand function must be postulated. Accomplishing this permits the parameters of the function to be estimated by statistical analysis (usually multivariate regression); and once the parameters are known, the price elasticity can be calculated.

Since explanatory variables are typically and sometimes strongly correlated with each other, it is helpful to collect data on every important explanatory variable, so that as many related factors as possible can be included in the multivariate models. Also, since some explanatory variables may not be identified (through oversight or lack of data), complete analysis requires considering the consequences of omitting relevant variables.

Demand models may be estimated from primary data (collected for the purpose, perhaps by means of specially installed meters) or from secondary data (collected for another purpose, usually billing). In all cases, water-use data are usually of moderate to poor quality. Observations are frequently missing, and reported observations may be incorrect. Secondary data may contain estimates of water use (where meter readings were not available), and the period covered by each water-use observation (the billing period) may be irregular.

The quality of observations of explanatory variables may also be a source of problems. In some cases, the variables are poorly specified: the defined variable may be similar to, but not the same as, the variable actually presumed to affect water use. For example, residential households differ in their capacity to use water because of differences in life-style and available water-using appliances; and since this variable cannot be measured directly, it is usually approximated by means of such variables as housing value, household income, number of appliances, educational attainment, and socioeconomic class. While each of these characteristics may capture some part of the relationship of interest, none is identical to the true explanatory variable.

As a result of missing or poorly specified data, most demand functions leave a significant fraction of the variance in water use unexplained. The problem is most noticeable in analyses based on cross-sectional data, where the missing variables are especially likely to affect the results.

## Price Elasticity of Demand Components

Boland et al. (1984) defined likely ranges of price elasticity for sectoral water use (winter, summer, and combined residential; commercial; and industrial). Price elasticity values were also divided, where ap-

**Table 3-1. Elasticity Range of Sectoral Water Use.**

| | Elasticity Range[1] | |
|---|---|---|
| Sector | Short-run | Long-run |
| Residential | | |
| winter | n.a. | 0.00 to $-0.10$ |
| summer | n.a. | $-0.50$ to $-0.60$[2] |
| sprinkling | n.a. | $-0.70$ to $-1.60$[3] |
| combined | 0.00 to $-0.30$ | $-0.20$ to $-0.40$ |
| Industrial | n.a. | $-0.30$ to $-6.71$[4] |
| Commercial | n.a. | $-0.20$ to $-1.40$[5] |

[1] Elasticity ranges defined as "most likely elasticity range" or "reported elasticity range" in the accompanying report.
[2] Elasticity range applies to eastern United States.
[3] Elasticity range applies to eastern and western United States.
[4] Elasticity range applies to individual categories of industrial water use. Elasticities for aggregate industrial use reported as $-0.50$ to $-0.80$.
[5] Elasticity range applies to individual categories of commercial water use.

SOURCE: Boland et al. (1984).

plicable, into a short-run category (with a time scale on the order of months to one year) and a long-run category (with a time scale on the order of several years or more). The results of the survey are summarized in table 3-1.

The reported values of price elasticity are subject to a number of statistical deficiencies that may lead to errors in elasticity estimates. These deficiencies originate in sample selection, model specification, choice of explanatory variables, choice of price variable, and level of aggregation. The survey findings are useful, however, in identifying the relative magnitude of elasticity among different sectors of water use and in identifying the relative magnitude of the range of elasticity within each sector.

## PRICE ELASTICITY AND WATER-INDUSTRY ECONOMICS

This section sets forth the reasons why water managers as a group consider the price elasticity of water demand to be insignificant (Hildebrand 1984).

## Water is Different

One of the basic beliefs of the water-utility industry is that water differs fundamentally from other goods or commodities, and therefore that demand cannot be predicted by conventional economic theory. This belief is noted in several studies and is criticized as being the cause of overestimating water demands and overdeveloping water supplies. Hanke (1970) claimed that,

> water policies are frequently based on misconceptions concerning the nature of the demand for residential water. The fundamental misconception that faces the water industry has been termed by Mulliman (1963) the "water-is-different" philosophy. This philosophy is reflected in a number of different ways, but in all cases it connotes that water is unique and should not be treated as an economic good.

Clearly, water is unique in at least two aspects: the massive amounts of it used, and its extremely low price.

Typical per-capita urban water use in the service area of the Metropolitan Water District of Southern California (MWDSC) is about 8,000 cubic feet or 250 tons per year. For southern California as a whole, some 87 billion cubic feet or 2.7 billion tons of water are imported annually, in addition to locally supplied water. Combining all sources of supply, 130 billion cubic feet or 4.0 billion tons of water are used annually in the MWDSC service area. But despite the massive quantities of water used, the price of water is extremely low. A typical retail water rate in southern California amounts to about $0.16 per ton of water (about 32 cubic feet) delivered to an individual residence, store, or factory. Delivered bottled water, on the other hand, costs about $6.00 per cubic foot or $192 per ton.

Wolman (1981) gives some interesting comparative cost figures for various commodities. Costs in the following list are expressed in dollars per acre-foot (1 acre-foot = 326,000 gallons, approximately the volume of water that would supply a family of five for one year, including lawn and garden irrigation):

| | |
|---|---|
| Water | $283 per acre-foot |
| Gasoline | $420,000 per acre-foot |
| Milk | $873,000 per acre-foot |
| Coca Cola | $1,668,000 per acre-foot |
| Beer | $1,419,000 per acre-foot |
| Scotch | $39,964,000 per acre-foot |

The low price of water is remarkable in view of the complex facilities required for its transport and storage, the ever-tightening quality criteria applied to it, and the universal delivery of it—which is rarely interrupted by drought, flood, or earthquake.

## Physical Facilities
The $0.16-per-ton commodity charge for delivered water in southern California includes the costs of aqueducts to transport the water from distant sources, of reservoirs to store it in order to guarantee that otherwise undependable supplies can meet variations in demand, and of systems to treat the water and distribute it to places of use. The massive quantities of water consumed effectively preclude it from being transported and distributed by train or truck, as most other commodities can be. Consequently, large and expensive facilities must be built to transport, store, treat, and distribute water.

It is generally impractical to build aqueducts or reservoirs in small increments to match year-to-year growth in water demand. Instead, such facilities are usually planned to serve forty to fifty years of future growth. Economies of scale dictate the large incremental jumps in capacity. In the case of distribution systems, capacity is usually provided for only twenty to twenty-five years of growth; even here, however, expensive facilities are constructed well in advance of need.

The large incremental additions to water-supply systems create a great deal of excess capacity that remains unused for many years. Ideally, a new facility is built when water demands are first projected to exceed the capacity of the system in the absence of the facility. Consequently, very little of the capacity is used initially, but increasing use is made as water demands grow with time. Eventually, as full-capacity use of the facility is approached, another large incremental addition is made to the water-supply system.

## Water-Industry Economics
Major incremental additions to a water-supply system require large capital expenditures. In order to distribute the cost of a new facility over a portion of its life, bonds are usually sold, and the revenues from them are used to pay construction costs. Nevertheless, a sizable increase in the total annual costs of the system occurs when such an

addition is made, and as a result the long-run average total cost curves for water utilities have discontinuities. The smooth, continuous curves of economic theory are not strictly applicable to water utilities.

## Effect of Demand on Water Price

Over a period of years, a water utility's revenues generally must equal its total costs, including reserve requirements. Reserve funds set aside in years of normal water sales can be used to pay fixed costs in years of deficient water sales; however, no utility could long survive on revenues that fall short of costs. Assuming, as is most often the case, that the utility is a nonprofit organization, revenues cannot exceed costs in the long run, either.

Water utility revenues may be derived from taxes, fixed charges, and water sales. (Minor sources of revenue such as income from investments and power sales, also exist.) For most water utilities, the bulk of the revenues needed to pay their total costs come from water sales. It follows that, for a group of utilities located in the same geographic area and having similar water-supply and distribution costs, the utilities having the lowest per-capita water sales will tend to have the highest water rates, and vice versa. It also follows that, for a given utility, a decrease in water sales in any year will tend to cause water rates to increase. Thus, a correlation between the price of water and the demand for water could be expected even if there were no price elasticity of demand for water.

## Uncertainties of Water Supply and Demand

Estimates of future water demand are made by utilities as guides in determining the need for and timing of additional water supplies. Yet most utilities encounter as much uncertainty in estimating future water supplies as in estimating future demands. These uncertainties and the consequences of underestimating or overestimating water demand are discussed in the following paragraphs.

**Uncertainty of Water Supply**   Aside from utilities that use large lakes (such as Lake Michigan), large carry-over storage reservoirs (such as the Colorado River system), or large groundwater aquifers (such as the Ogalla aquifer) as their source of water, most utilities are faced with uncertainties in their future water supplies. These uncertainties stem from the inability of forecasters to predict future weather conditions. Without very large amounts of storage to smooth out the year-to-year vagaries of rainfall and runoff, a potentially great degree of variation is possible in the water supply available in a given year. In response to this variation in supply, utilities commonly base their calculations of dependable supply on the amount of water that was available during a historical dry period—not necessarily the most adverse historical runoff period, but one estimated to recur infrequently enough to give a high degree of assurance that the supply recorded then would be available in the future. A supply that is exceeded in all but one year out of fifty (98-percent reliability) would be considered a dependable supply.

**Uncertainty of Demand**   Future water demands are difficult to determine because of two major uncertainties: future population growth, and future per-capita use. Per-capita water use heavily depends on local weather conditions, and years-in-advance long-range weather forecasting is simply not possible.

**Combined Probabilities**   Based on historical records of rainfall, temperature, and runoff, the probabilities that various water supplies will be available and that various levels of demand will occur in future years can be computed. These probabilities can be combined to determine the probability that water demands will be met in future years; and water supplies can then be developed to meet the area's water needs within any acceptable probability of future shortages. For example, if one year in fifty of having a water shortage is an acceptable risk, supplies can be so planned.

There is one catch in this otherwise logical methodology for planning water supplies: it is almost certain that, sometime in the future, a worse drought will occur than has occurred historically. Indeed, it would be most unusual if the short period of the available historical record should include the driest period possible in a region.

**Consequences of Incorrect Estimates**   The many uncertainties involved in estimating future water supplies and demands make it impossible to plan future supplies that will exactly meet demands. Because of this, the common practice of water utilities is to plan dependable supplies to meet dry-year demands. If water supplies smaller than the dependable supply or water demands greater than the dry-year demands (or both) should be encountered, water shortages will result.

On the other hand, if future water supplies should exceed those planned or if water demands should be less than estimated, an excess water supply will result for a few years, until the growth in demand equals the supply. When this situation arises, facilities that might have been deferred will have been built in advance of need. Traditionally, water planners have preferred erring on the side of providing facilities in advance of need to risking an increased probability and magnitude of water shortage.

Hildebrand (1984) suggests that, even if price elasticity has an effect on per-capita water use, the effect is small in comparison to other uncertainties that must be considered in making water-demand projections.

## REFERENCES

Boland, J. J.; Dziegielewski, B.; Baumann, D. B.; and Opitz, E. M. 1984. *Influence of Price and Rate Structures on Municipal and Industrial Water Use.* U.S. Army Corps of Engineers, Engineer Institute for Water Resources, Contract Report 84-C-2.

Craddock, E. 1985. Personal communication containing the draft information packet on water rates for the Office of Water Conservation, California Department of Water Resources, June 6, 1985.

Engineering News Record. 1986. Brakes on water rates. *Engineering News Record* 216(18): 5.

Goldstein, J. 1986. Full-cost water pricing. *Journal of the American Water Works Association* 78(2): 52–61.

Hanke, S. H. 1970. Demand for water under dynamic conditions. *Water Resources Research* 6(5): 1253–61.

Hildebrand, C. E. 1984. The relationship between urban water demand and the price of water. Report prepared for the Metropolitan Water District of Southern California, February 1984.

Orange County Municipal Water District. 1985. Urban Water Management Plan (draft), July 1985.

Russell, J. D. 1984. Seasonal and time-of-day pricing. *Journal of the American Water Works Association* 76(9): 63–65.

Wolman, A. 1981. Urban water supply: water resources Cinderella. *Journal of the American Water Works Association* 73(7): 28, 55.

# CHAPTER FOUR

# Water-Lifeline Hazard Mitigation

*Lifelines* are defined as services necessary for the survival of a community. Utilities included in this definition are potable water supply, sewage collection and treatment, electric power, natural gas, and communications systems. Transportation systems are also included in this definition, although they are not utilities in the ordinary sense.

Most water utilities are routinely subjected to hazards—either through supply-and-demand imbalances or through distribution problems. Pipes break, valves stick, hydrants crack, and power outages often occur. Some of the routine hazards faced by water utilities are described as follows (California Department of Water Resources 1985):

1. Supply-and-demand imbalance
   - Imbalance may result from abnormal demand during an unseasonable heatwave when supplies are already strained or from a sudden influx of visitors for a special event.
   - Processing and storage capacity may restrict the amount of water available to users, even though source supplies are adequate.
   - Equipment failures may prevent maximum use of existing water supplies.
   - Lowered water levels in wells may increase pumping requirements beyond the operating capacity of existing equipment.
2. Distribution problems
   - Fluctuating water-pressure levels may reduce the utility's ability to deliver water at a stable rate, creating or aggravating water-use problems.
   - Without an adequate backup capability in the system, a pipeline break may cause cessation of water delivery without warning.

Water utilities can also be subjected to extraordinary natural hazards such as hurricanes, floods, tornadoes, high winds, earthquakes, forest or brush fires, droughts, and tsunamis ("tidal waves" produced by submarine earth movement or volcanic eruption). Disasters such as extensive watershed fires can be extremely serious—as can blizzard conditions, which render inaccessible the appurtenances needed for management and control of the utility.

Allison and Shields (1985) classify water-resources hazards under two headings: natural hazards, such as storm, flood, and drought; and technical hazards, such as hazardous waste spills, computer failure, and toxicity. While natural hazards are to some extent expected and tend to be well-documented, relatively few technical hazards are recorded—especially in new and evolving technologies. In addition, the management of technical hazards is often limited by popular belief in the reliability or limited dangerousness of products of technology. Water lifelines are especially vulnerable to technical hazards because water supplies can be contaminated without any attendant damage to storage and conveyance facilities. Such contamination could result from seepage of chemicals into groundwater or surface supplies or from deliberate sabotage of the supply.

Various hazards, such as some instances of chemical contamination or spills, massive equipment failures, and explosions or incendiary fires, are caused by accident. Other hazards are man-made, including intentional contaminant-dumping, employee work stoppage, vandalism, and various civil disorders. Some hazards, such as a watershed fire, could be the result of a natural phenomenon, a careless accident, or a deliberate action to destroy property.

## WATER-LIFELINE HAZARDS

Each disaster or hazard has a specific effect on different parts of the water system (American Water Works Association 1980). Four subsystems of the water utility must be considered: collection, transmission, treatment, and distribution. Within each subsystem are critical components such as power, personnel, materials and supplies, and communications. Although each subsystem has its own particular strengths and vulnerabilities, damage to one part of the system may affect other parts of it. For example, a tornado is unlikely to harm pipes underground, but it could destroy all the power sources necessary for continuous operation of the remaining water system. A dam break or loss of

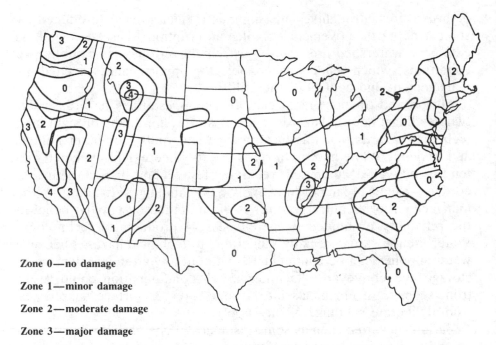

Zone 0—no damage

Zone 1—minor damage

Zone 2—moderate damage

Zone 3—major damage

Zone 4—extreme damage

Figure 4-1. Seismic risk zones in the continental United States. (Reprinted from *Emergency Planning for Water Utility Management*, p. 7, by permission; copyright © 1984, American Water Works Association)

reservoir water would not affect the pipes, pumping stations, or communications system, but losing the source of water would cause the system as a whole to close down.

## EARTHQUAKES

Earthquakes in the United States are identified with the Pacific coast: significant temblors have been recorded in Alaska and southern and central California. They are classified according to the level of damage they cause and according to their probability of occurrence. The level of damage earthquakes can cause is shown in figure 4-1. The seismic map delineates damage zones that are marked with numbers ranging from 0 to 4; zone 0 represents no damage, and zone 4 represents extreme damage.

**Table 4-1. Earthquake Energy Relationships.**

| Magnitude[1] | Expected Annual Incidence[2] | Distance Felt[3] | | Area Felt | | Energy Released[4] |
|---|---|---|---|---|---|---|
| | | (mi) | (km) | (sq mi) | (sq km) | (tons TNT) |
| 3.0–3.9 | 49,000 | 15 | 24 | 0.75 | 1.9 | ~10 |
| 4.0–4.9 | 6,200 | 30 | 48 | 3 | 7.8 | 14–200 |
| 5.0–5.9 | 800 | 70 | 113 | 15 | 39 | 230–10K |
| 6.0–6.9 | 120 | 125 | 201 | 50 | 129 | 14K–200K |
| 7.0–7.9 | 18 | 250 | 403 | 200 | 520 | 230K–10M |
| 8.0–8.9 | 1 | 450 | 723 | 800 | 2070 | 14M–200M |

[1] Richter scale.
[2] Gutenberg, B., and Richter, C. F. 1954. *Seismicity of the Earth and Associated Phenomena.* Princeton, N.J.: Princeton University Press. 1954.
[3] Benioff, H., and Gutenberg, B. 1955. General Introduction to Seismology, Earthquakes in Kern County During 1952. California Division of Mines, Bull. 171.
[4] K = 1,000; M = 1,000,000.

SOURCE: Reprinted from *Emergency Planning for Water Utility Management*, by permission; copyright © 1984, American Water Works Association.

New risk maps developed by the Federal Emergency Management Agency divide the country into seven zones, based on the probability of earthquake occurrence (*Engineering News Record* 1986). The new maps give design-acceleration levels that have a 90-percent probability of not being exceeded in fifty years. Previously, risk maps were based on the magnitude of past earthquakes but not on their frequency. Combining both factors provides a more realistic picture of earthquake activity in the United States and can be used in the development of new seismic safety codes.

The degree of damage an earthquake causes is related to its magnitude (table 4-1) and to the location of the property with respect to the *epicenter* (the part of the earth's surface directly above the focus of an earthquake). In the case of a water utility, disaster effects include disrupting pipelines, damaging dams, damaging or destroying powerlines (and thereby isolating water utilities), destroying water-intake systems, breaking well casings, and physically destroying components of water-treatment plants. A more detailed discussion of seismic hazard mitigation is given later in this chapter.

## HURRICANES

Hurricanes in the United States are identified with the eastern seaboard and the Gulf coast. The six-month period from June 1 to November 30 is considered the Atlantic hurricane season, but hurricanes usually occur in August, September, and October. Hurricanes are generally classified according to the damage they produce. The classification ranges from 1 to 8, with class 8 storms causing the least amount of damage. Major hurricanes are relatively rare events at any location. An average of six Atlantic hurricanes occur per year.

Hurricane-force winds and accompanying high tides can have a devastating effect on shoreline areas. Disaster effects include damage to surface structures from the winds, power outages, flooding of plant facilities, and extensive contamination of water supplies.

## TORNADOES

Tornadoes in the United States are identified with the Great Plains states. A tornado is a difficult natural phenomenon to describe, but its destructive power resembles that of a hurricane. With the possible exception of earthquakes, tornadoes occur with less warning than any other natural disaster, and they may result in high loss of life.

## FLOODS

Floods are a potential threat wherever there is a major river basin or a shoreline along an ocean or sea. Floods are the easiest type of natural disaster to predict and to defend against, provided that sufficient funds are available at a national and regional level to pay for preventive measures. Disaster effects of flooding include contamination of water supplies, inundation of treatment facilities and pumping stations, power outages, disruption of transportation and telephone facilities, and scattering of stockpiled equipment.

## TSUNAMIS

Tsunamis are generated by undersea earthquakes; in the United States, they are identified with the Pacific coast and the state of Hawaii. About twenty-five earthquakes with magnitudes exceeding 5.0 on the Richter scale occur in the oceans each year, undetected by the continental network of seismic stations (*Water Newsletter* 1986). These undersea quakes are detected by an array of eleven hydrophones set up on the ocean floor near Wake Island in the center of the Western Pacific. It is the only seismic network in the region and covers an area about as large as the entire continent of North America. Scientists monitoring the Wake Island array speculate that the thin ocean crust on which the array is placed may prevent seismic waves from traveling far into the continents. Scientists believe that the quakes detected in this area may indicate the presence of a previously unrecognized tsunami-generating zone within the Pacific basin.

Disaster effects of tsunamis are the same as those of flooding, but they are compounded by the tremendous quantities of energy contained in the tidal wave that accompanies the undersea earthquake. For example, the tsunami caused by a 1964 earthquake in Alaska completely destroyed entire communities and generated unusually high tides as far south as La Jolla, California.

## RIOTS, VANDALISM, CIVIL DISORDER, WORK STOPPAGES

Collectively, these people-induced actions can cause any or all of the following:

- Contamination of water supplies
- Large-scale waste of water
- Disruption of pumping facilities
- Loss of power
- Disruption of communications
- Interruption of maintenance and repair operations
- Curtailment of services
- Destruction of watershed areas
- Physical harm to the plant and to maintenance personnel
- Destruction of property
- Disruption of water treatment by limiting availability of chemicals, supplies, and the like

## HAZARDOUS MATERIAL SPILLS/DUMPING

Spilling and dumping of hazardous materials can originate from pipelines, boats, airplanes, motor vehicles, railroads, or fixed containers. The materials released may be liquid, powdered, or gaseous material. Their release can be accidental or intentional, and it can occur without warning at any time of day or night. For example, in 1981, a crop-dusting plane carrying herbicides crashed into a central California river not far from the water intake to a city's water supply. Disaster effects of spillage and dumping include contamination of water supplies, loss of life, and damage to property. The cost to the water system and to water users affected by the spill could be very high—not only for cleanup, but for replacing components of the system and for defending against lawsuits for damages to those affected.

## COMPUTER FAILURE

Causes of computer failure include electrical problems (such as loss of power), component failures, software failures (such as destruction of files by accident), telephone-line failures, air-conditioning problems, and deliberate tampering. Adverse effects that computer failure might cause include having to operate filter plants, pump stations, and valves manually; delivering bills late; processing payments slowly; losing records; and having to rebuild files.

## TOXICITY

*Toxicity hazards* arise when chemical and/or biological agents are added to a water supply in such a concentration and at such a location as to cause immediate illness or death to people who ingest the contaminant. The degree of hazard depends on the concentration required to cause injury, the speed with which the action takes place, and the degree of severity of the action. Table 4-2 defines toxic agents that represent acute potential threats to a water utility.

## SUMMARY

Table 4-3 summarizes the relationship between specific disasters and their potential effects. Even if a disaster of one of these types directly injured only a small component of a water utility, the entire water system could be affected because of the interdependence of its components.

Every water utility in the United States, regardless of its size and geographic location, is subject to the effects of several types of disasters.

**Table 4-2. Potential Acute Toxic Agents in
Drinking Water.**

| | Concentration | |
|---|---|---|
| Agent | $mg^1$ | ppm |
| LSD[2] | 0.005 | 0.1 |
| Botulinus toxin[3] | 0.001 | 0.002 |
| Staphylococcus enterotoxin[4] | 0.05 | 0.1 |
| Nerve agents[4] | 50 | 100 |
| Arsenic[5] | 100–300 | 200–260 |
| Cyanide[4] | 25 | 50 |
| Fluoride (sodium fluoride)[4] | 3000 | 6000 |
| Cadmium[5] | 15 | 30 |
| Mercury[5] | 75–300 | 150–600 |
| Dieldrin[6] | 5000 | 10,000 |

[1] Based on the ingestion of 500 ml (16 oz) of water.
[2] *Journal of the American Water Works Association* (Jan. 1967); 120–22.
[3] WHO (1970).
[4] Bell, F. A. Letter report (April 1972).
[5] McKee, J. E., and Wolf, H. W. 1963. Water Quality Criteria. California State Water Quality Control Board, Pub. No. 3-A.
[6] DuBois, K. P. 1959. Pesticides, Rodenticides, Herbicides, Household Hazards. Information Circular on Toxicity of Pesticides to Man, WHO, No. 2.

SOURCE: Reprinted from *Emergency Planning for Water Utility Management*, by permission; copyright © 1984, American Water Works Association.

Still, disasters of various types significantly differ in their likelihood and potential severity. The energy release associated with earthquakes and hurricanes overshadows that of other natural disasters, and the tremendous energy of floods has been shown to be significantly controlled by long-term, well-engineered flood-control projects. It is evident that flood warnings give some preparedness advantage, but the suddenness of an earthquake, tornado, civil disorder, or chemical spill precludes immediate predisaster action. If disasters are thought of together with their effects, however, as shown in table 4-4, disaster effects are much less identifiable with a specific disaster than most people would expect.

Certain disasters, such as earthquakes or hurricanes, cause far more widespread harm than do more localized occurrences—a riot, strike,

**Table 4-3. Interrelationship of Disasters and Their Effects.**

| Effect | Hurricane | Tornado | Bomb Blast | Earthquakes | Flood | Explosion of Cargo Ships, Oil Tanks, etc. | Plane Crash | Civil Disorder | Epidemic | Volcanic Dust | Industrial Discharge | Spills of Hazardous Materials |
|---|---|---|---|---|---|---|---|---|---|---|---|---|
| Structural damage | • | • | • | • | • | • | • | • | | | | |
| Water/sewer lines broken | • | • | • | • | • | | | | | | | |
| Storage tanks destroyed or contaminated | • | • | • | • | • | | | | | | | |
| Power lines down | • | • | • | • | • | | | | | | | |
| Access limited by debris | • | • | • | • | • | | | | | | | |
| Fires | | | | | | • | • | • | | | | |
| Water pollution | | | | | | | | | | • | • | • |
| Power outages | | | | | | • | • | • | | | | |
| Disease | | | | | | | | | • | | | |
| Air pollution | | | | | | | | | | • | • | • |
| Land contamination | | | | | | | | | | • | • | • |
| Taxing of chlorination facilities | | | | | | | | | | | | |

SOURCE: Reprinted from *Emergency Planning for Water Utility Management*, by permission; copyright © 1984, American Water Works Association.

**Table 4-4. Disaster Effects Matrix.**

| Type of Disaster | Plant (Construction) Damage | Watershed Damage | Reservoir Damage | Storage Tank Damage | Broken Mains | Contamination | Communication Disruption | Transportation Failure | Employee Shortages |
|---|---|---|---|---|---|---|---|---|---|
| Earthquake | • |  | • | • | • | • | • | • | • |
| Hurricane | • |  | • | • | • | • | • | • | • |
| Flood | • |  | • |  | • | • | • | • | • |
| Tornado | • |  | • | • |  | • | • | • |  |
| Tsunami |  | • | • | • |  | • | • | • |  |
| Riots | • | • | • | • | • | • | • | • | • |
| Spills | • | • | • |  |  | • | • | • | • |

SOURCE: Reprinted from *Emergency Planning for Water Utility Management.* by permission; copyright © 1984, American Water Works Association.

tornado, or dumping incident. But the effects of these events, in many cases, differ only in scale. For example, measures designed to offer a higher degree of operational reliability during a hurricane will have the same advantage in case of a tornado or flood.

## ECONOMIC IMPACT OF HAZARD MITIGATION

Determining the economic impact of hazards to water lifelines carries the same degree of uncertainty as does predicting the hazard itself. Water supplies often are supplied through regional networks that encompass large areas and cross city and county boundaries. Hence, the economic impact of losing this supply should be considered from both regional and local perspectives.

### REGIONAL PERSPECTIVE

The most prevalent way of expressing economic loss from natural disasters is in terms of property damage. A *major disaster* is usually defined as a disaster that has caused property damage greater than some arbitrary threshold amount such as $100 million and/or has exceeded some casualty threshold. This method has two drawbacks: it conjures an image of an instantaneous, one-time effect; and it underassesses the value of the more important flow of services obtained from property and assets.

A study by the Department of Economics of the University of California at Riverside suggests that the value of forgone production of goods and services constitutes a superior measure of loss (Rose 1980). Output losses and related employment losses offer a basis of comparison that applies a more solidly grounded threshold of concern. Recessions in the United States, which are usually caused by the vagaries of the business cycle, by inept fiscal and monetary policies, and by outside influences such as concerted increases in international oil prices, have typically resulted in output decreases of 2.0 to 4.0 percent (over at least a six-month span) since World War II. There is no reason why a hazard-induced recession at the regional level should warrant any less concern.

Table 4-5 illustrates the frequency of significant economic downturns caused by natural hazards, and it relates these to the hazard type and

**Table 4-5. Frequency of Natural Disasters.**

| Type | Scope of the Economy Affected (typical cases) | 2% Output Loss[1] Threshold (no. per year) |
|---|---|---|
| Drought | Regional | 0.20 |
| Earthquake | Local/Regional | 0.35 |
| Tsunami | Regional | 0.10 |
| Flood (regular) | Local/Regional | 2.20 |
| Flood (flash) | Local | 0.10 |
| Forest fire | Local/Regional | 2.80 |
| Hurricane | Regional | 1.05 |
| Tornado | Local/Regional | 2.60 |

[1] Period of coverage: 1955–75 in most cases.

SOURCE: Adapted from Rose (1981), p. 109, by permission of ASCE.

the scope of the economy affected. The frequency of natural hazards is the figure for the whole nation; frequencies for subregions are much lower.

## LOCAL PERSPECTIVE

A study by the Federal Emergency Management Agency provides an interesting insight into the problem of hazard mitigation from a utility's viewpoint (Thiel 1981). In the event that a hazard occurs and a utility lifeline is damaged, funds for repair are most likely to come from the following sources:

**1.** For publicly owned utilities, the President's Disaster Fund pays 100 percent of the repair and upgrading costs, if the President declares a Federal Disaster. This is almost certain to occur in response to any disaster that inflicts even modest damage.
**2.** Some costs are borne directly by the local taxpayers.
**3.** For private utilities, the rate structure is adjusted to recoup repair costs not covered by grants-in-aid.

In the aftermath of a disaster large enough to damage utility lifelines, a utility manager should not find it difficult to defend the rate or tax

increases required to repair the system. A rational manager may con-
clude, therefore, that the system is already fully insured. In this way,
economic analysis yields the astounding conclusion that little or no
hazard mitigation should be undertaken beyond current practices.

## CONCLUSIONS

In any lifeline engineering problem, a utility manager is likely to
propose an investment that pays off only if a hazard actually occurs.
This type of problem is an example of decision-making under uncertainty.

The traditional approach of water utilities to hazard mitigation has
been to provide a sufficient amount of emergency storage to withstand
a long-duration outage of primary water-supply lines. Little investigation,
however, is done of the economic consequences of this approach or
of the probabilities of actually incurring an outage of sufficient duration
to cause hardships. Hazard mitigation in any form usually costs something, and hence the importance attached to it must be weighed against
other needs.

Social and economic considerations should be included in any
decision-making process involving hazard mitigation for utility lifelines.
Utility bills for several types of lifelines have increased dramatically
in recent years, and further increases are likely to be resented—
especially in instances where the additional service (for example,
insurance against loss of service and all that ensues) is not obvious
to the average customer. Many studies by utilities and responsible
agencies indicate that, in the long run, the social and economic costs
of hazard-induced disruption to utility lifelines is probably less than
the social and economic costs that would be imposed on society to
enhance the lifelines' resistance beyond current levels.

## HAZARDS AND PUBLIC POLICY

In general, major lifelines and associated critical facilities are clustered
in geographic areas that contain high population concentrations. Major
lifeline systems generally must cross various geologic environments
and be exposed to various natural and man-made hazards.

A UCLA study found that, while there is considerable public fatalism
about earthquakes and other natural hazards, many people believe

that steps can and should be taken on behalf of individuals who are especially endangered by these hazards (Turner 1979). This study indicates that people look overwhelmingly to local and regional governmental and regulatory bodies to take these steps.

Natural hazards can be dealt with in either of two ways: steps can be taken to avoid disasters in advance, or measures can be developed to deal with their effects in the aftermath. Therefore, the problems posed by natural hazards require more than conventional technological solutions. In many cases, governmental bodies, such as local water utilities, deal with hazard mitigation by influencing the behavior of the individuals, groups, and communities living at risk from these hazards. In this respect, hazard mitigation must be inherently political and technological.

Public policy is a statement of objectives, actions, and directives for achieving a stated goal; it is expressed in laws, regulations, and executive orders, and it is enforced by regulatory agencies. Components of public policy include jurisdiction, comprehensiveness, and risk management. Federal, state, or local agencies may have regulatory jurisdiction over specific facility types. Policy comprehensiveness may be measured against requirements for siting, design, review of design, inspection of construction, and operations of a facility. Risk management is reflected by the policy approach, examples of which include emergency preparedness, land-use planning, and engineering solutions; various combinations of these alternatives are also possible.

A comprehensive study of state and local politics with respect to natural hazards, covering twenty states and one hundred local communities, concluded that natural hazards are not very high on the agenda of things that people worry about (Wright 1981). Furthermore, the study indicated that most people will accept planned solutions to natural hazards only if the economic and political costs are commensurate with the seriousness of the problem as it appears to them. Such an attitude was also prevalent in a survey conducted in Orange County, California; only 27 percent of those surveyed considered earthquakes to be a serious problem (*Orange County Register* 1984). Of greater concern to residents were water supply (54 percent), air pollution (52 percent), and hazardous waste (43 percent).

On the subject of water resources, there is increasing public pressure toward establishing a no-risk environment where government, employers, and social organizations play key roles in risk-elimination efforts (Aharoni 1981). Basically, people are unwilling to accept risk in many water-resources decision-making efforts, once the risks are known and judged. Some risk may be accepted if government assurances are offered in the form of low-cost insurance or a history of substantial government interaction through grants, loans, and other instruments of rehabilitation (Allison and Shields 1985).

## WATER LIFELINES

Water lifelines consist of two distinct systems: water-storage facilities (reservoirs), and water-distribution systems (major aqueducts, major distribution networks, and local distribution networks). Public policy regarding dams and reservoirs is highly developed on both the federal level and the state level; it has been directly tied to failures and near-failures of dams, including the failure of the St. Francis Dam in southern California in 1928, the failure of the Baldwin Hills Dam in Los Angeles in 1962, and the near-failure of the Lower San Fernando Dam in Los Angeles during the 1971 San Fernando earthquake.

## WASTEWATER SYSTEMS

Relatively little public policy has been formulated to address risk reduction for wastewater systems. No federal or state legislation is directly concerned with this issue. Water-quality and water-pollution-control legislation exists at both federal and state levels. Although comprehensive performance requirements exist for wastewater-treatment facilities, including specifications for acceptable quality levels of discharge effluent, the possible consequences of treatment-plant failure as a result of earthquake or other disaster-induced damage are usually not addressed.

For wastewater-collection systems, policies establishing standards for pipeline design and location, as well as other aspects of the system, may be adopted by local agencies. If these facilities are the responsibility of a special district, the city or county may not be involved at all. Consideration given to natural hazards and to the implementation of risk-reduction measures varies greatly from district to district.

## RISK-REDUCTION PROGRAMS

The importance of the objectives of lifelines—to maintain public health and safety, and to ensure the continued operation of critical facilities—

indicates that measures for risk reduction should be enforced through the development and implementation of risk-reduction programs, as part of public policy.

As described previously, public policy to reduce risks for lifeline facilities ranges from comprehensive to nonexistent. The comprehensive policies developed for water-storage reservoirs are based on the evaluation of risks. Failure of a dam would result in inundation of the area below the dam; if an urban area is located in that area, extensive loss of life and property damage may occur. The amount of risk acceptable to the public for dam failure is thus very low. Regulatory agencies' policies reflect this low level of acceptable risk by establishing stringent design and operation requirements, and by requiring mapping of the potential area of inundation for use in establishing emergency-response plans. Policies for most water-distribution systems, wastewater-treatment systems, and wastewater-collection systems, however, are very limited or do not exist. Some public or private lifeline operators may independently develop and implement risk-reduction measures; others simply may not address the risks.

Developing a system of risk classification may facilitate understanding of risks and the development of public policy for risk reduction. *Calculated risk* is the estimated amount of risk for a given facility. *Acceptable risk* is the risk that can be accommodated without undue hardship. *Residual risk* is the difference between the calculated risk and the acceptable risk; in other words, it is the amount of risk that needs to be managed. With these definitions, a comprehensive risk-reduction program can be undertaken in the following steps (Patwardhan et al. 1981):

1. Identify the calculated risk associated with natural and man-made hazards.
2. Estimate the residual risk, based on value judgments of acceptable risk.
3. Implement risk-reduction measures through the development of public policy.

**Step 1: Identify Calculated Risk.** An understanding of the amount of loss (the calculated risk) that may be sustained by a facility as a result of foreseen or unforeseen disasters is of utmost importance. The estimate should include valuations for both primary physical loss and secondary consequential loss (such as loss of water supply for fire-fighting). Calculated risk is directly related to the nature of the facility and to the exposure or susceptibility of the area to hazards, both natural and man-made. The performance of individual facilities and the probable occurrence of a range of losses should be estimated.

Risk analysis, as it applies to hazard mitigation, begins with projecting water-supply shortages of variable severity and duration. This projection can be obtained by simulating monthly water use at a future point in time, using the forecasting methods and techniques discussed in chapter 2. Projected monthly demands are then subtracted from the historical monthly available supply to arrive at simulated monthly surpluses or deficits. Next, historical records are analyzed to establish the probabilities of different deficit situations, involving varying degrees of intensity and lengths of duration. Figure 4-2 outlines a similar procedure used by Boland, Carver, and Flynn (1980), in which estimates of the probability of drought occurrences were based on forty-five years of historical data.

An economic loss is assigned to each water-shortage situation, based on an appropriate economic indicator. Calculated risk is then defined as the expected value of the loss, and it equals the sum of the products of the probabilities and their associated losses (Young, Taylor, and Hanks 1972).

**Step 2: Estimate Acceptable and Residual Risk.** Acceptable risk for a particular facility should now be assessed. This assessment involves a value judgment about the importance of the facility and the potential consequences of its failure. Factors that may influence acceptable risk include the size of the population that would be affected, the duration of the effect, the degree of difficulty or cost associated with repairs, and the degree of system interaction or redundancy in the area.

Evaluating calculated and acceptable risks allows an assessment of residual risk. Table 4-6 provides an example of possible levels of residual risk (expressed as reliability levels) for water and wastewater lifelines under varying degrees of seismic loading. This table is excerpted from a table proposed by Duke (1981). System reliability is illustrated for two levels of intensity of vibratory motion and surface faulting, using the modified Mercalli scale; the calculations apply to new and

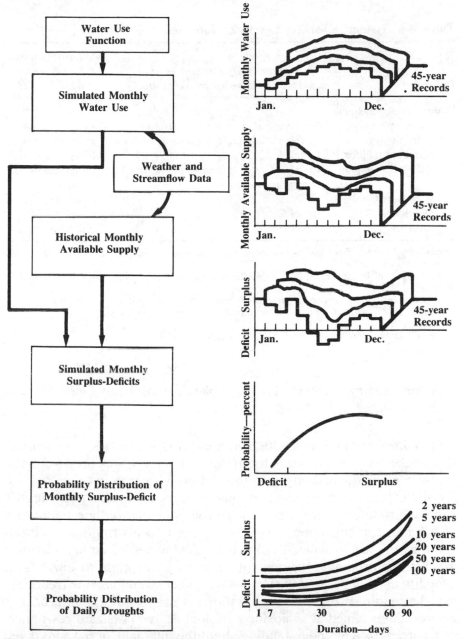

Figure 4-2. Steps in predicting water deficits. (Reprinted from *Journal of the American Water Works Association* 72(7): 369, by permission; copyright © 1980, American Water Works Association)

**Table 4-6. System Reliability Levels for Lifelines.**

| Lifeline | High-intensity Ground Motion (MMI IX–X with surface faulting) | Moderate-intensity Ground Motion (MMI VI–VII) |
|---|---|---|
| Water Supply System | | |
| storage reservoirs | No failure that will endanger lives | Fully functional |
| for fire fighting | Level A[1]: adequate storage available | Fully functional |
| treatment facilities | Level A[1] | Fully functional |
| distribution system | Level B[2]: tank trucks available for potable water | Fully functional |
| Sewage System | | |
| collection | Level C[3] | Level A[1] |
| treatment | Level B[2] | Fully functional |

[1] Reliability level A: 5 percent or less of intensity area is without service for one day; fully restored in one week.
[2] Reliability level B: 20 percent or less of intensity area is without service for one week; fully restored in one month.
[3] Reliability level C: 50 percent or less of intensity area is without service for one week; 20 percent or less is without service for one month; fully restored in 3 months.

SOURCE: Adapted from Duke (1981), p. 8, by permission of ASCE.

expanded existing facilities. For an explanation of the modified Mercalli scale, see tables 4-8 and 4-9.

**Step 3: Implement Risk-Reduction Measures.** The risk-management approaches most appropriate for lifelines are emergency preparedness, land-use planning, and engineering solutions. Emergency preparedness for lifelines may involve developing emergency plans that identify sources of materials for emergency repairs or that outline temporary distribution or collection systems or alternate distribution methods (such as the use of water trucks) in critical areas. Land-use planning involves controlling development in areas susceptible to hazards; if a facility must be located in such an area, special engineering treatment may be required. Engineering solutions consist of design approaches or procedures aimed at achieving a desired degree of hazard resistance. The current lack of understanding about the full range of risks involved may limit the value of engineering solutions for some types of facilities.

The ultimate goal of a comprehensive plan for risk reduction is to

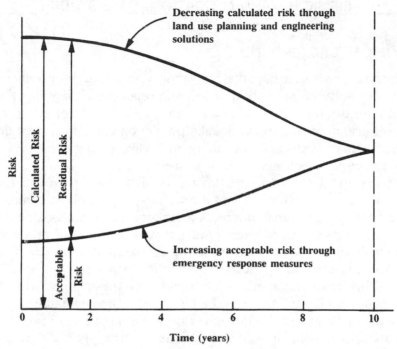

Figure 4-3. Conceptual planning diagram for reducing seismic risk for lifelines. (Reprinted from Patwardhan et al. 1981, p. 30, by permission of ASCE)

eliminate residual risk. Since residual risk is equal to calculated risk minus acceptable risk, the difference can only be narrowed by decreasing the calculated risk or increasing the acceptable risk. Risk-reduction measures are used to accomplish this. For example, emergency preparedness may be used to increase the acceptable risk. Potentially adverse impacts on the population as a whole can be diminished by developing alternative means of providing a service. Advanced preparation can ensure more effective response to repair damages, minimizing the time of service interruption. Calculated risks can sometimes be lowered by implementing land-use planning measures that limit the location of existing lifelines in areas subject to flood inundation, fault rupture, or ground failure. A conceptual model of a seismic risk–reduction plan implemented over a ten-year period is shown in figure 4-3.

## SEISMIC HAZARD MITIGATION: CASE STUDIES

## HISTORICAL BACKGROUND

The problem with earthquake-risk analyses is that they depend on how accurately the prediction of seismic recurrence identifies timing, magnitude, duration, and location (Jahnus 1979). Unfortunately, the seismological approach to earthquake prediction contains a large degree of uncertainty. Historical seismicity provides a very short record of a phenomenon which takes place in a geologic time frame.

The historical record of seismicity is all too short in the United States and in many other parts of the world—especially in comparison to the average recurrence intervals for major earthquakes along given faults. Findings by a Cal Tech geologist reveal that at least nine very large earthquakes have occurred along the south-central reach of the San Andreas fault in California since the sixth century (Seih 1971; Seih 1978). The average return period between these large earthquakes is estimated to be 160 years at Pallet Creek (approximately 20 miles southeast of Palmdale, California). For some large active faults, such as the Garlock fault in southern California, the historical record is simply a gap between the most recent earthquake (which must have occurred in very early prehistoric times) and the next event, which has not yet taken place.

Consequently, seismologists normally work on the conservative side of extrapolation, from what little data are available. Even then, the seismologist may be concerned that the available data may represent information taken during a period of soon-to-end quiescence. Some scientists have made statistical analyses of worldwide seismicity in order to enlarge local data bases for projections into the future, recognizing that such projections can only be generally applied to a given fault or a given region.

### Earthquake Loss Data: Worldwide
During the past 4,000 years, approximately 13 million lives have been claimed by earthquakes worldwide. About 3.5 million of these deaths have occurred during the past 400 years, and nearly 2 million during the past century. These figures do not include the Mexico earthquakes that struck on September 19 and 20, 1985, with Richter-scale magnitudes of 8.1 and 7.5, respectively (*ASCE News* 1985), the El Salvador earth-

quake that occurred on October 10, 1986 with a Richter-scale magnitude of 5.4, nor recent minor earthquakes.

Thus, the rate of earthquake-related fatalities has been increasing, partly because of world population growth and partly because of increasing concentrations of people in certain vulnerable areas of the world. There have, however, been many departures from this broad trend, and the occurrence of a single great earthquake in or near a metropolitan area can cause the statistics to change dramatically. In 1976, for example, although fewer than the average annual number of large earthquakes occurred worldwide, about 700,000 people were killed in earthquakes—more than 600,000 of them in the Tangshan, China, event. The heavy losses from this one earthquake drastically raised the global fatality total for this century.

## Earthquake Loss Data: California and the United States

Total earthquake-related fatalities in this country have amounted to only about 1,700 since the beginning of the nineteenth century. Approximately two-thirds of that total can be assigned to California temblors, and nearly one-half to a single California event—the San Francisco earthquake of 1906. More than one hundred damaging events have occurred in California since 1800, but seventeen of these have accounted for virtually the entire earthquake-related death toll; these seventeen quakes have caused an average of 65 fatalities per event, with the events occurring at an average rate of one per decade. Table 4-7 outlines the complete history of recorded earthquakes in California. Tables 4-8 and 4-9 explain earthquake intensity levels and their relationship to the Richter scale.

The averages mentioned in the preceding paragraph are not very meaningful, because the fatalities are unevenly distributed among the earthquakes and because the earthquakes are unevenly spaced in time. For example, if the 1906 San Francisco earthquake is excluded from consideration, the average death rate from the sixteen other fatal California earthquakes since 1800 drops from 65 per event to less than 20 per event. The skewing effects related to concentrations of people are further evidenced by a drop to 4 deaths per event if four other earthquakes are excluded: 1812 Capistrano, 1869 Hayward, 1933 Long Beach, and 1971 San Fernando.

During the period from 1925 to 1975, 590 deaths in the United States have been attributed to earthquakes. This figure amounts to only about

## Table 4-7. Historical Earthquakes in California.

| Year | Date | Place | Modified Mercalli Intensity | Remarks |
|------|------|-------|------------------------------|---------|
| 1769 | July 28 | San Pedro Channel area | X | Major disturbances with many aftershocks |
| 1790 | Unknown | Owens Valley | X | Major shock with appearance of fault scraps |
| 1812 | Dec. 8 | San Juan Capistrano | IX | Church collapsed killing 40 |
| 1812 | Dec. 21 | Near Lompoc | X | Churches and other buildings wrecked in several towns, including Santa Barbara |
| 1836 | June 10 | San Francisco Bay area | X | Ground breakage along Hayward Fault from Mission San Jose to San Pablo |
| 1838 | June | San Francisco | X | Fault rupture phenomenon along San Andreas rift, this earthquake is probably comparable to the earthquake of April 18, 1906 |
| 1857 | January 9 | Fort Tejon | X–XI | One of the greatest historical Pacific coast shocks; originated on San Andreas fault in northwest corner of L.A. county; buildings and large trees thrown down |
| 1865 | October 8 | San Francisco peninsula | IX | Extensive damage in San Francisco, especially on filled ground |
| 1868 | October 21 | Hayward | X | Many buildings wrecked and damage in Hayward and East Bay; severe damage at San Leandro and San Francisco; 30 killed; rupture of Hayward fault |
| 1872 | March 26 | Owens Valley | X–XI | One of the greatest earthquakes in Pacific Coast area; 7-meter fault scrap formed; 27 killed at Lone Pine out of 300 population; adobe houses wrecked |
| 1899 | December 25 | San Jacinto | IX | Nearly all brick buildings badly damaged in San Jacinto and Hermit; chimneys down in Riverside; 6 killed; another severe shock in 1918 |

| Year | Date | Place | Modified Mercalli Intensity | Remarks |
|------|------|-------|-----------------------------|---------|
| 1906 | April 18 | San Francisco | XI | Great earthquake and fire; about 80 percent of estimated $400 million damage due to fire; 700 killed; greatest destruction in San Francisco and Santa Rosa; horizontal slipping along San Andreas fault, 6.5 meters; greatest damage on poorly filled land |
| 1915 | June 22 | Imperial Valley | VIII | Nearly $1 million damage; 6 killed; well-constructed buildings cracked |
| 1925 | June 29 | Santa Barbara | IX | $6 million damage; 13 killed; 70 buildings condemned |
| 1933 | March 10 | Long Beach | IX | $41 million damage; 120 killed; fire damage insignificant |
| 1940 | May 18 | Imperial Valley | X | $6 million damage; 8 killed, 20 seriously injured; 65-kilometer fault appeared with maximum horizontal displacement of 4.5 meters |
| 1941 | June 30 | Santa Barbara | VIII | $100,000 damage |
| 1941 | November 14 | Torrance, Gardena | VIII | About $1 million damage; 50 buildings severely damaged |
| 1952 | July 20 | Kern County | X | $60 million damage; 12 killed, 18 seriously injured; railroad tunnels collapsed and rails bent in S shape; surface faulting with about 0.5 meter of vertical, as well as lateral, displacement |
| 1952 | August 22 | Bakersfield | VIII | 2 killed, 35 injured; damage $10 million |
| 1957 | March 22 | San Francisco | VIII | Damage in Westlake and Daly City area |
| 1969 | October 1 | Santa Rosa | VII–VIII | Property loss of $6 million; felt over an area of 30,000 sq. kilometers |
| 1971 | February 9 | San Fernando | VIII–XI | $500 million direct physical loss; 65 killed, more than 1,000 injured; felt over an area of 230,000 kilometers |

SOURCE: U.S. National Oceanic and Atmospheric Administration.

**Table 4-8. Approximate Relationships between the Intensity Scale and the Richter Scale.**

| Richter Magnitude | Expected Annual Incidence Worldwide[1] | Distance Felt (stat. mi.)[2] | Intensity (Maximum Expected Modified Mercalli)[3] |
|---|---|---|---|
| 3.0–3.9 | 49,000 | 15 | II–III |
| 4.0–4.9 | 6,200 | 30 | IV–V |
| 5.0–5.9 | 800 | 70 | VI–VII |
| 6.0–6.9 | 120 | 125 | VII–VIII |
| 7.0–7.9 | 18 | 250 | IX–X |
| 8.0–8.9 | 1 | 450 | XI–XII |

[1] Gutenberg, B., and Richter, C. F. 1954. *Seismicity of the Earth and Associated Phenomena.* Princeton, N.J.: Princeton University Press.
[2] Benioff, H., and Gutenberg, B. General Introduction to Seismology, Earthquakes in Kern County During 1952. California Division of Mines, Bull. 171.
[3] U.S. Department of Defense, 1962. *Effects of Nuclear Weapons*, ed. S. Glasstone. Washington, D.C.: Government Printing Office.

3 percent of the fatalities from floods, hurricanes, tornadoes, and earthquakes during the same half-century period. The worldwide percentage for earthquake deaths in comparison to deaths attributable to other natural disasters is four times as high as the corresponding figure for the United States, owing to differences in seismicity levels, topographic and climatic conditions, population concentrations, lifestyles, construction practices, and other factors. Earthquake-related deaths in the United States have amounted to about 15 percent of the death toll from hurricanes, and to less than 10 percent of the death tolls from floods and from tornadoes. They have been at about the same level as fatalities from avalanches and from lightning strikes. In fact, the fatalities from all natural disasters in the United States during recent decades have amounted to only about 2 percent of those from vehicular traffic.

Cumulative physical damage from earthquakes in the United States has been estimated at approximately $2 billion since 1800. This total seems small in comparison to the $4 billion price tag for a single nonseismic event, Hurricane Agnes in 1972. The average cost of 10 cents per person for earthquake damage during the past century also seems small in comparison to the average cost of $25 per person for all natural disasters in the United States during the same period.

**Table 4-9. Modified Mercalli Intensity Scale.**

| Intensity Value | Description[1] |
|---|---|
| I | Not felt. Marginal and long-period effects of large earthquakes. |
| II | Felt by persons at rest, on upper floors, or favorably placed. |
| III | Felt indoors. Hanging objects swing. Vibrations like passing of light trucks. Duration estimated. May not be recognized as an earthquake. |
| IV | Hanging objects swing. Vibration like passing of heavy truck; or sensation of jolt like a heavy ball striking the walls. Standing cars rock. Windows, dishes, doors rattle. Glasses clink. Crockery clashes. In the upper range of IV, wooden walls and frames creak. |
| V | Felt outdoors; direction estimated. Sleepers wakened. Liquids disturbed, some spilled. Small unstable objects displaced or upset. Doors swing, close, open. Shutters, pictures move. Pendulum clocks stop, start, change rate. |
| VI | Felt by all. Many frightened and run outdoors. Persons walk unsteadily. Windows, dishes, glassware broken. Knick-knacks, books, and so on fall off shelves. Pictures off walls. Furniture moved or overturned. Weak plaster and masonry D[1] cracked. Small bells ring (church, school). Trees, bushes shake visibly, or can be heard to rustle. |
| VII | Difficult to stand. Noticed by drivers. Hanging objects quiver. Furniture broken. Damage to masonry D, including cracks. Weak chimneys broken at roofline. Fall of plaster, loose bricks, stones, tiles, cornices, also unbraced parapets and architectural ornaments. Some cracks in masonry C.[1] Waves on ponds, water turbid with mud. Small slides and caving in along sand or gravel banks. Large bells ring. Concrete irrigation ditches damaged. |
| VIII | Steering of cars affected. Damage to masonry C; partial collapse. Some damage to masonry B[1]; none to masonry A.[1] Fall of stucco and some masonry walls. Twisting, fall of chimneys, factory stacks, monuments, towers, elevated tanks. Frame houses move on foundations if not bolted down; loose panel walls thrown out. Decayed piling broken off. Branches broken from trees. Changes in flow or temperature of springs and wells. Cracks on wet ground and on steep slopes. |
| IX | General panic. Masonry D destroyed; masonry C heavily damaged, sometimes with complete collapse; masonry B seriously damaged. General damage to foundations. Frame structures, if not bolted, shifted off foundations. Frames racked. Serious damage to reservoirs. Underground pipes broken. Conspicuous cracks in ground. In alluviated areas, sand and mud ejected, earthquake fountains, sand craters. |
| X | Most masonry and frame structures destroyed with their foundations. Some well-built wooden structures and bridges destroyed. Serious damage to dams, dikes, embankments. Large landslides. Water |

*(continued)*

**Table 4-9.** (*continued*)

| Intensity Value | Description[1] |
|---|---|
| | thrown on banks of canals, rivers, lakes, and so on. Sand and mud shifted horizontally on beaches and flat land. Rails bent slightly. |
| XI | Rails bent greatly. Underground pipelines completely out of service. |
| XII | Damage nearly total. Large rock masses displaced. Lines of sight and level distorted. Objects thrown into the air. |

[1] Masonry A, B, C, D. To avoid ambiguity of language, the quality of masonry—brick or otherwise—is specified by the following lettering:
*Masonry A:* Good workmanship, mortar, and design; reinforced, especially laterally, and bound together by using steel, concrete, and so on, designed to resist lateral forces.
*Masonry B:* Good workmanship and mortar; reinforced, but not designed in detail to resist lateral forces.
*Masonry C:* Ordinary workmanship and mortar; no extreme weaknesses like failing to tie in at corners, but neither reinforced nor designed against horizontal forces.
*Masonry D:* Weak materials, such as adobe; poor mortar; low standards of workmanship; weak horizontally.

SOURCE: Richter, C. F. 1958. *Elementary Seismology.* New York: W. H. Freeman and Company. Copyright © 1958.

Four-fifths of the earthquake costs were associated with three events: 1906 San Francisco, 1964 Alaska, and 1971 San Fernando. However, $1 billion in direct damage and indirect costs were incurred from the San Fernando event—a moderate earthquake located in a metropolitan area. The California Division of Mines and Geology considered physical and demographic factors in predicting a startling $21 billion in earthquake damage for the state during the period 1970–2000, assuming a continuation of present responses to earthquake hazards.

Water supplies in the western United States are especially prone to seismic hazards. While each destructive earthquake has led to some improvement in antiseismic design for lifelines, certain California shocks have had greater influence than others. The following review of the lifeline impacts of these selected earthquakes identifies actions that were taken afterward to help reduce losses from subsequent earthquakes (Duke 1981):

1. *San Francisco, 1906.* This experience led to improvements in water-supply systems, principally for fighting fire. The importance of system redundancy was emphasized.

**2.** *Long Beach, 1933.* The temblor here resulted in general adoption of lateral-force provisions in California building codes. Several California electrical utilities adopted antiseismic design criteria for important facilities, exceeding those required by local building codes.
**3.** *Kern County, 1952.* Electrical utilities improved their design standards for anchoring and bracing electrical equipment. Some earthquake-resistant features were developed for tanks located on the ground, and the importance of flexibility in connected piping was emphasized.
**4.** *San Fernando, 1971.* This shock has had a broad impact on the development of lifeline earthquake-resistant design. Systems with and without antiseismic design features were subjected to strong ground-shaking and to differential earth movements. The behavior of lifelines in this earthquake indicated some glaring weaknesses.

The 1971 earthquake in San Fernando caused people to focus more attention on utility lifelines than they had done in the past. Prominent damage was done to high-cost facilities such as public and private hospitals and utility lifelines—including freeways, sewage-treatment facilities, water-distribution systems, key components of electrical energy systems, telephone equipment, and a major dam. Thus, the San Fernando earthquake brought into focus the vulnerability of lifelines and critical facilities to earthquakes, and it inspired numerous efforts to assess and improve the situation.

The record of United States earthquakes and their effects has supported conflicting theories. Some experts regard seismic hazards as relatively minor threats in comparison to other physical hazards. Large-magnitude earthquakes in this country are infrequent, and the costs associated with historic events have been relatively low. Thus, it can be argued that dealing with earthquake risks should have relatively low priority in the allocation of a utility's time and resources.

A conflicting theory is premised on the idea that fatalities and property damage are influenced by the magnitude and duration of a strong earthquake, as well as by the number of people living in the area and the nature of their buildings and other structures. The recorded history of earthquakes in the United States includes only one great event that occurred in a heavily populated region—San Francisco in 1906. Unsurprisingly, this earthquake has dominated the statistical record of social losses; but substantially higher losses must be anticipated when another great earthquake occurs in a metropolitan area of the United States. Moreover, just such an occurrence is highly likely during the next hundred years.

## CASE STUDIES

The following case studies illustrate how some utilities operating at risk from natural hazards have incorporated concepts of risk evaluation and risk reduction into long-range planning for hazard mitigation.

### Pacific Gas and Electric Company

A study by the Pacific Gas and Electric Company sheds light on the role of economics in seismic hazard mitigation (Bettinger 1980). This study identifies two principal benefits that can result from seismic-hazard mitigation by a utility system: reduced economic losses from damage in the event of an earthquake; and greater capability to provide continuous service in the event of an earthquake. Assuming that some reasonable degree of seismic resistance has already been incorporated into a utility system that anticipates minor damage to but no gross failure of its major elements, the more important of the two benefits probably is continuity of service.

The PG&E report states that the cost of damage expectable in a large earthquake is probably insufficient to justify the cost of system-wide modifications to guarantee damage-free survival in any specific area. The damage, though heavy in some areas, might be more easily paid for by investing the money and having the profits available to pay for the damage when, where, and if it occurs. The report recommends that any utility contemplating major expenditures for seismic-hazard mitigation first determine whether the expenditure is one that will benefit the rate-payer in the long run. No utility is so vulnerable that it will suffer widespread outages at the slightest earth tremor; neither is any utility wealthy enough to spend huge sums of money on a system designed to avoid the slightest consumer inconvenience in the event of an 8.2-magnitude earthquake like the one San Francisco experienced in 1906.

Since continuity of service concerns the consumer, the vulnerability of utility distribution systems to earthquake hazards is most often the hazard cited in justifying expenditures for seismic-hazard mitigation. Distribution systems have suffered damage in earthquakes, but the damage normally is of a type that is relatively easy to repair. As a practical matter, if damage to the distribution system is severe, considerable damage to customers' facilities is also extremely probable. The fact that utilities have made some allowance for seismic loads in their design means that service is likely to be restored before many customers are able to accept and use it.

Although the damage incurred by a distribution system in a large earthquake may be relatively easy to repair, it will probably be widespread. Thus, many crews with equipment will be needed. In this regard, utility systems covering large geographic areas have a built-in capability for coping with disasters because they can quickly bring in crews and equipment from unaffected areas. Most utilities also have standing arrangements with neighboring utilities to allow additional repair teams to converge rapidly on troubled areas. For example, Southern California Edison reports that, of the 250,000 electricity customers who lost service during the 1971 San Fernando earthquake, 180,000 suffered an interruption of less than one minute, while another 30,000 had service restored within an hour (Department of Commerce 1973).

## Utah

The Utah Seismic Safety Advisory Council is charged with developing state policies for mitigating earthquake hazards. In a preliminary study, the council found that formulating such a policy entails more than organizing current technical information about earthquakes, system response, and mitigation techniques (Ward and Taylor 1980). The council concluded that technological advances constitute only one essential part of the process by which seismic mitigation measures are implemented for utility lifelines. The process also requires that legal and political realities be considered.

Of particular importance to water utilities is the council's finding that individual components of utility systems have varying degrees of vulnerability and importance. A system analysis that combines damage estimates and assessments of system operation is needed to determine the importance of any given component's ability to withstand an earthquake. Table 4-10 shows the approximate damage thresholds for various water-supply system components. For an explanation of earthquake intensity levels and their relationship to the Richter scale, see tables 4-8 and 4-9.

**Table 4-10. Comparative Predicted Earthquake Effects on Components of Water Supply Systems in Utah.**

| Component | Approximate Earthquake Intensity Threshold Level for Damage |
|---|:---:|
| Pump station (unreinforced masonry building) | VII |
| Poorly sited tank | VII |
| Underground pipes in poor condition | VII |
| Underground pipes (cast iron or asbestos cement) | VIII |
| Inlet and outlet connections to reservoirs | VIII |
| Contamination of wells | VIII |
| Temporary power outage | VIII |
| Unanchored equipment (transformers, tanks, pumps, and so on) | VIII |
| Outlet connections to treatment plants | IX |
| Unanchored tanks | IX |
| Large pipes at fault crossings | IX |
| Underground pipes (welded steel, some ductile iron, some pvc) | IX |
| Buried reservoirs | X |
| Anchored tanks | X |

SOURCE: Adapted from Ward and Taylor (1980), p. 211, by permission of ASCE.

## EBMUD

The East Bay Municipal Utility District (EBMUD) is located on the east side of San Francisco Bay. Two major active faults cross its service area, two active faults cross its supply aqueducts, one active fault passes through its principal water-supply reservoir, and its service area is subject to major ground-shaking from the San Andreas fault.

EBMUD has taken a number of actions to minimize seismic damage and to maximize recovery of its water facilities for purposes of fighting fires and maintaining at least a base level of water service (Anton 1980). Four broad lines of action have been pursued:

- Preparing and organizing for rapid and extensive internal repairs
- Making immediate low-cost/high-potential operational preparations or modifications to existing facilities
- Improving planning and design standards to reflect latest seismic experience
- Reviewing the seismic adequacy of existing critical operating fa-

cilities on a priority basis, with seismic strengthening undertaken or planned as needs are identified

Salient features of these four action plans are outlined in the paragraphs that follow.

**Planning for Internal Repairs**  The first broad line of action taken by EBMUD was to plan for a strong internal repair capability to ensure rapid restoration of service after an earthquake—primarily by utilizing its own repair stock, equipment, and manpower. EBMUD has undertaken cooperative actions with the State Office of Emergency Services and other utilities; these include periodic review and testing of communications, review of repair supplies, and rolling stock inventories. Other forms of cooperation range from stockpiling quick-coupling pipe to water-tank trucks that can set up emergency water points. Emergency interconnections with neighboring utilities have been installed on a shared-cost basis.

To restore water supply to dry areas, water can be brought in by carefully opening upper-zone valves after broken mains have been isolated. From lower zones, water can be pumped by portable pump units or firetruck pumpers that have been connected to fire hydrants separated by zone line valves.

**Advance Preparations and Modifications to Existing Facilities**  EBMUD facilities have been designed to bypass chlorinated raw water at all filter plants in case a plant, its equipment, or its processes should fail. Reasonable quantities of water-treatment chemicals are stockpiled to ensure continued functioning of filter plants after an earthquake, in case chemical deliveries are curtailed. Critical operating facilities are equipped with alternative electrical and auxiliary power systems. At all filter plants, auxiliary diesel-powered generators automatically come on-line when a power failure occurs.

**Improved Planning, Data-Base, and Design Standards**  Seismic overlay maps are maintained for EBMUD's distribution system, to show the latest understanding of active and inactive fault lines and other information critical to water-system facilities. Geologic and soils-engineering studies are conducted at all new sites to avoid building in vulnerable areas.

**Review of Seismic Adequacy and Strengthening**  Critical pipelines in fault zones have been redesigned to include such features as above-ground lines, installation in conduits, flexible joints, double-welded joints, and additional valves. Double-welded steel pipe joints are used in critical areas and extra valves are installed. In addition, a 4-mile-long transmission main is planned for the southern portion of the distribution system, to parallel a section of a major aqueduct that crosses or lies within an active fault zone.

## DROUGHT MANAGEMENT

The incidence of drought has increased in this century, as concentrations of human settlement and food production have approached or exceeded the capacities of their natural resource base. Woo (1982) defines drought as a pervasive hazard because it develops gradually. The amount of precipitation, stream flow, or groundwater deficiency that constitutes a drought in one region may not necessarily have this status in another. The criteria for drought are the type and amount of water needed and the alternative supplies available to the region within a specified period of time. Drought occurrence in this century has increased in both impact and frequency. The years 1956 through 1971 are recognized by meterologists as an unusually stable period in terms of both temperature and precipitation, indicating that the probability of drought may be higher in the future.

## HISTORICAL BACKGROUND

Droughts do not strike with the suddenness and ferocity of earthquakes, but their effects can be even more profound. The following examples illustrate actions water utilities have taken in the face of a drought and unexpected problems they have encountered.

During a drought in 1969 in Sao Paulo, Brazil, unusual numbers of school children were seemingly afflicted by diarrhea, which forced them to spend considerable time in the bathrooms, using and flushing the toilets (Jezler 1980). An investigation revealed that, by wasting the water, the children had hoped to have classes canceled and be promoted to the next grade level by decree (without examinations), all because of the water shortage. The situation was quickly resolved when a law was passed forbidding closing schools, since a fleet of trucks was out delivering water. Soon after the law appeared, the students' intestinal troubles quickly disappeared.

On the island of Okinawa in 1971, the water supplier adopted a simple and effective method of dealing with recurrent water shortages: it completely shut the water off (Garland 1980)! The length of time during which water was shut off depended on the severity of the shortage. In 1971, the water was on for twelve hours and off for thirty-six hours throughout Okinawa's hot, humid summer. During the winter of 1973–74, the water was shut off for eight hours and turned on for sixteen hours until January, when the hours were extended to twenty-four hours on and twenty-four hours off. In 1963, the water shortage was so critical that temporary pipelines were laid from the ocean to inland distribution points, and seawater was used for nonpotable purposes.

In Great Britain, during a drought in 1976, public water supplies to 1 million people in southern Wales were cut off seventeen hours per day for up to eleven weeks (Blackburn 1980). In North Devon, where distribution to individual customers was shut down completely, 65,000 consumers took buckets of water home from standpipes erected in the streets by local authorities. A pressure reduction of 25 percent was introduced throughout the metropolitan distribution system and produced a 10 percent savings in water use. By implementing this means of reducing demand, the utility was able to maintain distribution throughout the system, but houses on the outer fringes had water only on their first floors.

Even when the rains began in September of that year, the situation was far from settled. The runoff from London's rainfall drained into an estuary that lay well below the water-supply intakes. Only upstream rains that contributed to water in the storage reservoirs helped replenish London's supply. London residents found it difficult to understand why the rain pouring down on them did not end their water shortage. Heavy rains also fell on upstream communities that depended on wells, but the wells filled slowly. Supplying aquifers did not recharge because the parched soil was soaking up everything that fell on it. In North Devon, citizens found themselves in the curious position of having to stand in drenching rain in order to obtain water for their houses from standpipes.

In California, during the drought of 1976–77, most communities embarked on voluntary and mandatory conservation programs (Robie 1980; Harnett 1980; Griffith 1980). Some of the rationing programs that were adopted allowed households as little as 50 gallons per day (gpd). In major population centers—the Monterey Peninsula communities, the East Bay Municipal Utility District, and the Marin Municipal Water District—high conservation rates were achieved, with cutbacks of 49 percent, 39 percent, and 53 percent, respectively, for the first nine months of 1977 over levels for the same period in 1976. One result of the cutbacks was that a number of water utilities had to adjust water rates upward to compensate for the decline in consumption. Several utilities used lifeline rate structures to work out inequities to the poor, elderly, and disadvantaged.

During prolonged droughts, water utilities often resort to drastic measures to curtail water consumption and prevent water waste. In Corpus Christi, Texas, a severe multiyear drought beginning in 1982 forced a local utility to fine customers $200 for lawn-watering. Limited watering of trees, shrubs, and some perennials was permitted on specific days only (American Water Works Association 1984). In Denver, Colorado, in 1977, the Denver Water Board imposed a 30 percent reduction in the annual number of new customers that would be added to the water system. This tap-allocation program was followed by a mandatory water-conservation program that limited outside watering to a maximum of three hours every third day (Miller 1980). In the midst of a severe water shortage, New York City passed regulations prohibiting some outdoor water uses and enforced them by using police helicopters to spot filled swimming pools and lawns that looked "suspiciously" green (*U.S. Water News* 1985a). In addition, New York City waiters and waitresses were subject to $100 fines for serving water in a restaurant to anyone who had not requested it (*U.S. Water News* 1985b). In the summer of 1980, a rationing program in New Jersey limited individuals to 65 gallons of water per day and families to 50 gallons per person, per day, under the threat of a $175 fine or one-year prison term (Hamilton 1984).

## DROUGHT CRITERIA AND INDICATORS

Before the effects of drought can be mitigated, its onset must be identified (Woo 1982). Long-term weather prediction is not yet accurate and has been hindered by recently increased variability in the weather.

**Table 4-11. Palmer Index Values and a State's Drought Stages.**

| Palmer Index | Drought Stage | Palmer Index | Drought Stage |
|---|---|---|---|
| 0.49 to −0.49 | Normal | −3.00 to −3.99 | Warning |
| −0.50 to −0.99 | Normal | −4.00 to −4.99 | Emergency |
| −1.00 to −1.99 | Alert | −5.00 or less | Disaster |
| −2.00 to −2.99 | Alert | | |

SOURCE: U.S. Army Corps of Engineers (1983a), p. B-26.

Concern has shifted to the establishment of drought indices to establish values for drought thesholds. Indices include: precipitation measurement; stream flow; reservoir, natural surface, and groundwater storage; soil moisture; temperature; and area geologic and geographic characteristics.

The most popular indicator of meteorological drought, the Palmer Index, was developed as a means of summarizing and periodically disseminating drought and crop-moisture information regionally. The index uses historical records of temperature and precipitation to develop five constants (pertaining to evaporation, recharge, runoff, moisture loss, and precipitation) that define the moisture characteristics of a given region.

A survey of water-management programs conducted by the U.S. Army Corps of Engineers (1983a) cited an application of the Palmer Index tied to response levels in a drought-management plan. The relationship between the index and five drought stages included in the management plan are shown in table 4-11. Under normal conditions, drought-monitoring and appraisal capacities are maintained, and contingency plans are kept up-to-date. The initial drought stages (alert and warning) activate monitoring and appraisal of drought conditions, and promote voluntary water conservation. The later stages (emergency and disaster) require implementation of mandatory water-use restrictions, use of emergency sources and equipment as necessary, initiation of actions to meet worst-case situations, and application to the federal government for disaster assistance.

One state identified in the Army survey bases its drought assessments on four hydrologic indicators (precipitation, reservoir and lake storage, stream flow, and groundwater levels), which are weighted on a regional basis and used in conjunction with the Palmer Index. The indicators are as follows:

**1.** *Precipitation.* Cumulative departure of precipitation over a certain time period is a measure of drought severity.

**2.** *Reservoir/lake storage.* Several criteria may be used for reservoir and lake storage to determine drought status:

- Current storage as a percentage of usable capacity, compared with historic or normal storage
- Number of days of water supply remaining
- Inches or rain required to fill a reservoir

**3.** *Stream flow.* Stream flow has two components: base flow, made up of discharges from groundwater; and surface runoff, resulting directly from precipitation. Monthly frequency curves are used, since they are the readings most applicable to drought determinations.

**4.** *Groundwater levels.* Observation wells in uplands and basin floors are used to monitor shifts in water availability in underground aquifers. Based on long-term records, maximum, minimum, and average groundwater levels have been established and can be compared with present levels to determine drought status.

## TRADITIONAL APPROACH TO DROUGHT MANAGEMENT

Traditionally, drought management of water-supply sources has been determined in accordance with the concepts of *safe yield* and *reliability of supply*. *Safe yield* is often understood as that output of a water-supply project that can be maintained during a severe drought, such as the worst drought in the historic record. The estimated probability that such a drought will occur in any given year defines the *reliability of supply* of the safe yield; for example, it may be 1/40 or 2.5 percent. A planner calculates this probability of failure by selecting a *design drought*—a drought with a stated probability of occurrence, such as 1/20 (5 percent), 1/50 (2 percent), or 1/100 (1 percent).

Selecting the design drought is the most important decision in the whole planning process, since it implicitly establishes the magnitude of economic losses that may be incurred in the long run. After specifying a design drought—for example, a 1-in-100-years drought—the planner assumes the required safe yield of the project to equal the level of demand at the end of the planning horizon. In the case of staged projects, timing and sizes of additions to source capacity may be found by applying the criterion of a minimum present value of all costs incurred in meeting the demand at all times (except during droughts that are more severe than the design drought).

Planning for short-term supply deficiencies is constrained by the physical dimensions of a water-supply system. The safe yield of the system is of limited value to a water planner. Although the system is designed to supply water at the specified safe yield during the design drought, the planner never knows how long the drought will last and how severe it will be. In the case of reservoir storage, maintaining the safe yield throughout an indefinitely long drought would just empty the reservoir, provided that the ongoing drought were as severe as the design drought. The responsibilities of the water planner during a drought force the adoption of a more sophisticated strategy, so that the well-being of consumers is protected by something more reliable than simple adherence to the safe yield. The short-term alternatives, however, are usually limited to a choice between temporary reductions in water use and utilization of emergency water supplies.

To keep the risk of running out of water at a reasonably low level, the water planner must always try to adjust the level of withdrawal to existing conditions during the ongoing drought—and especially to the actual volume of water in storage. In most cases, increasingly severe water-use restrictions are imposed, each keyed to water levels in the reservoir so that the critical level of storage is never reached. In other words, the objective during an actual water shortage is to reduce the level of supply as early as possible, in order to avoid the consequences of more severe cutbacks in water delivery at later stages of the drought.

## PERSPECTIVES ON CONTINGENCY PLANNING FOR DROUGHTS

While establishing drought thresholds helps in identifying the onset of a drought period, the measures taken to counteract potential drought effects (for example, water conservation) can be vitiated by a problem characteristic of some natural hazards: the lack of public recognition of the hazard (Woo 1982). For instance, residents of arid areas are more attuned to the possibility of drought than are residents of humid areas. Urban dwellers in either area, however, are less aware of the hazard of drought than are their rural counterparts. Further, residents of humid areas (especially urban dwellers) take more risks in terms of failing to prepare for drought.

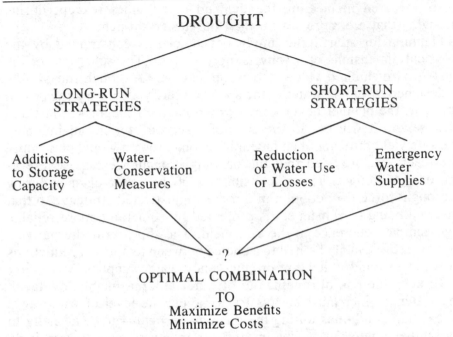

Figure 4-4. Two dimensions of drought contingency planning. (Reprinted from *Evaluation of Drought Management Measures for Municipal and Industrial Water Supply*, p. 3, U.S. Army Corps of Engineers, Engineer Institute for Water Resources, Contract Report 83-C-3, December 1983)

A comprehensive approach to evaluating drought contingency measures calls for determining the best mix of short-term and long-term strategies to deal with a range of projected water deficits. Short-term water strategies include reducing water use (or losses) and using emergency water supplies. Long-term strategies include adding storage capacity and instituting water-conservation measures. Figure 4-4 illustrates these two dimensions of contingency planning for droughts.

For the existing capacity of a water-supply system, the range of possible supply deficits in each future year is determined by comparing system yields with water-use forecasts. A set of various volumes of deficit—each corresponding to the probability of, say, 20 percent, 1 percent, 5 percent, 2 percent, and 1 percent—can be defined for each year. A minimum-cost emergency plan for coping with each deficit is determined by selecting an optional mix of short-term and long-term strategies, producing a set of cost estimates. Each estimate is then assigned a probability of occurrence, so that the higher costs associated with large volumes of deficit have a lower probability of occurrence.

The process of formulating contingency plans for droughts should also be evaluated from the long-term perspective of planning for water-supply capacity. The need to expand supply capacity or to implement long-term water-conservation programs should be assessed in conjunction with drought contingency plans. Long-term investment strategies for providing adequate urban water supplies should be evaluated according to the principles of planning for effective use of limited resources. The term *adequate* implies the principle of balancing the cost of supply additions and/or long-term water-conservation programs against the damages anticipatable from recurrent droughts in the long run. This problem is complicated by unrestricted demands for water that grow over time. In determining the timing and sizes of additions to the supply system, the water planner must identify a level of damages to expect from temporary shortages of water in the long run. The damages that actually result from a water shortage event, however, are difficult to estimate precisely unless intensive data-gathering is undertaken.

Suppliers have traditionally objected to demand management during contingency planning for droughts, or other water-supply shortages (Woo 1982). This objection is based on the belief that demand management requires suppliers to become allocators instead of suppliers. They argue that their mandate is to supply water according to demand. Managing demand before developing all supply options leaves suppliers open to the charge of mismanagement. On the other hand, suppliers may be equally open to the charge of mismanagement if they allow the supply of water to be exhausted. Their mandate is not simply to deliver water, but to deliver it from an assured supply. Viewed in this light, the role of the supplier must include management of both supply and demand.

## DATE REQUIREMENTS

Contingency planning for droughts requires well-quantified information in three areas: climatologic or engineering indicators that signal the onset of a drought and allow planners to estimate the risk of a supply shortage at each point of an actual drought event; the expected impacts of drought on the quantity and the quality of water in supply sources; and the magnitude of increased drought-period demands for water. Substantial amounts of data are required on water use in and socio-economic characteristics of the service area. These data are needed for developing the residential water-demand model and the disaggregate water-use forecast. The actual kinds and sources of site-specific data needed are shown in table 4-12. Data requirements for developing drought contingency plans are summarized in the following paragraphs.

In order to determine the magnitude of the supply deficit well-quantified information on climatologic or engineering indicators are needed. In order to invoke a drought contingency plan, a water-supply manager has to rely on a predetermined drought-alert procedure. Most often such procedures are based on arbitrarily established symptoms displayed by the water-supply system, such as the level of water in a reservoir, a river stage, or the depth of water in wells. Although they are very easy to monitor, these indicators cannot provide an early alert; on the other hand, they may reduce the number of false alarms (and consequent premature implementation of drought-management actions), thus avoiding various unnecessary costs on the system.

One way to improve the reliability of drought forecasting is by estimating (in terms of probability) potential drought conditions related to public water supplies in a larger region, such as the entire state. Once the regional drought is recognized, systems that are vulnerable to a drought-related water shortage may carry out a more detailed analysis of the risk of water-supply shortage in their areas. The important question to a water planner is whether or not the existing reserves of water are sufficient to support the system throughout the drought. To answer this question, the planner must determine the magnitude of the expected supply deficit. Several techniques that can be used to produce such estimates are summarized in table 4-13.

In selecting an optimum drought-management plan, the water planner must consider alternatives for augmenting available supplies. Emergency water supplies are defined as auxiliary sources that can provide limited-quantity or lower-quality water during periods of water shortage; table 4-14 lists the most common types of such sources. Each emergency water supply should be evaluated in the following terms:

• Availability and quality of water in potential auxiliary sources during long-term dry-weather conditions

**Table 4-12. Types and Sources of Data.**

| Data Category | Specific Item | Source |
|---|---|---|
| Operational | Disaggregated historical water use; daily total volumes of water treated and delivered; daily reservoir levels; water and sewer rate structure and revenues for each customer category | Water utility; state, regional, or local planning agencies |
| Water resource utilization | Water-use statistics for other than municipal users located in the area (self-supplied industry, agricultural users)—especially if they use the same source | Water utility; interviews with individual users; planning agencies |
| Hydrographic and hydrologic | Topographic maps; reservoir rating curves; drainage area(s); streamflow and raw-water-quality records; hydrologic characteristics of alternative sources, including yield estimates and water quality and minimum flow requirements | U.S. Geological Survey; Environmental Protection Agency; planning agencies |
| Climate | Precipitation; air temperature; snowfall; soil moisture conditions | National Weather Service (NOAA) |
| Socioeconomic | Population, household size and income; number of households and housing units; value and size of residential properties; number of commercial and institutional establishments, and value of receipts | U.S. census of population housing, business; U.S. Bureau of Labor Statistics |
| Land-use | Fractions of land in various use categories (such as urbanized, cropland, woodland); agricultural production statistics | U.S. census of agriculture; planning agencies |
| Costs | Investment and operation-maintenance costs for alternative water-supply sources; operation and maintenance cost of water-supply system | Water utility; consultant engineering reports; U.S. Environmental Protection Agency reports; U.S. Army Corps of Engineers; planning agencies |
| Psychological/ legal | Acceptability of demand reduction measures by customers; legal/ institutional obstacles | Literature studies; local government; water utility; interviews of individual customers |

SOURCE: U.S. Army Corps of Engineers (1983c), p. 5.

**Table 4-13. Methods for Forecasting the Availability of Water Supply.**

| Procedure | Products | Required Input Data |
|---|---|---|
| Basin Climatic Index (BCI) method | Expected total for 12 months' runoff, with 10, 25, and 50 percent probability of occurrence | Drainage basin or regional data: long-term average BCIs and runoff, monthly precipitation, and temperature |
| Position analysis | Percent probability of complete exhaustion of the reservoir storage during drought | Monthly inflow, withdrawals, and evaporation for a reservoir plus current reservoir storage |
| USGS technique | Percent probability of a dry reservoir based on representative trace of inflows | Historical and filled-in stream-flow data |
| National Weather Service river forecasting systems | Simulated stream flows; total volume of flow; maximum, minimum, and average mean daily flow | Hydrological parameters and initial conditions of a watershed, including moisture storage contents, snow-pack water-equivalents, future time-series of mean areal precipitation, and temperature (at least 10–20 years of record) |
| Snow accumulation and ablation model | Snow cover outflow plus rain that fell on bare ground | Air temperature, snow-pack water-equivalents, other snow-cover variables |
| Sacramento soil moisture accounting | Five components of water flow: direct runoff; surface runoff; lateral drainage interflow; supplementary baseflow; primary baseflow | As for the NWS RFS model |
| Sensitivity approach (for the NWS RFS rainfall-runoff procedures) | As for the NWS RFS model | Typical trace of 6-hour-interval rain data, current soil moisture, variance of rainfall input |
| Stochastic conceptual hydrologic model (based on NWS RFS system) | Stream-flow forecasts 6, 12, 18, 24, 30, and 36 hours in advance | Rainfall data in 6-hour time steps and incoming real-time discharge |

SOURCE: U.S. Army Corps of Engineers (1983b), pp. 26–27.

**Table 4-14. Illustrative List of
Emergency Supplies.**

Interdistrict Transfers
  · Interconnections: emergency; neighboring
    community; new connection
  · Importation: by trucks; by railroad cars

Cross-purpose Diversions
  · Alternative uses: hydropower; flood
    control; recreation
  · Stream flow: minimum flow requirements;
    recharge; downstream users

Auxiliary Sources
  · Surface water: untapped creeks, ponds,
    and quarries; dead reservoir storage;
    temporary pipeline to a river
  · Groundwater: abandoned wells; new wells

Cloud Seeding

SOURCE: U.S. Army Corps of Engineers (1983*b*), p. 60.

• Adequacy of existing treatment facilities to produce finished water
of acceptable quality when emergency supplies make up some fraction
of the new water supply
• Lead time required to construct necessary water transmission and
pretreatment facilities (if required)
• Construction and operation/maintenance costs entailed in bringing
emergency sources on line
• Potential obstacles to implementation, such as institutional barriers,
right-of-way considerations, and operational permits

In addition to forecasts of total municipal demand, short-term disaggregate water-use forecasts may be required, in order to evaluate
the effectiveness of water-conservation measures. Since these water-
conservation measures affect different classes of water use in different
ways, their effectiveness cannot be determined from aggregate data.
Estimates of drought damages, too, should be prepared for specific
categories of customers. For purposes of drought contingency planning,
disaggregated time-series water-use data should be used, if available.

## REFERENCES

Aharoni, Y. 1981. *The No-Risk Society*. Chatham, N.J.: Chatham House.

Allison, R. C., and Shields, J. 1985. Risk taking and risk making in water resource management. Paper presented at the ASCE spring convention in Denver, Colorado.

American Water Works Association. 1980. *Emergency Planning for Water Utility Management*. Manual M19. Denver: AWWA.

American Water Works Association. 1984. Texas lawn-watering brings $200 fine. *Mainstream* 28(8): 3.

Anton, W. F. 1981. Water lifeline earthquake engineering practiced by EBMUD. In *Social and Economic Impact of Earthquakes on Utility Lifelines. Proceedings of the ASCE Construction Division Specialty Conference in San Francisco, California, May 1980*, ed. J. Isenberg, pp. 125–33. New York: ASCE.

ASCE News. 1985. Report of Technical Council on Lifeline Earthquake Engineering. *ASCE News* (December 1985): 11.

Bettinger, R. V. 1981. Economics and seismic hazard mitigation for a gas and electric utility (Pacific Gas and Electric Company). In *Social and Economic Impact of Earthquakes on Utility Lifelines. Proceedings of the ASCE Construction Division Specialty Conference in San Francisco, California, May 1980*, ed. J. Isenberg, pp. 98–106. New York: ASCE.

Blackburn, A. M. 1980. Management strategies: dealing with drought. In *Water Conservation Strategies*, pp. 16–24. Denver: AWWA.

Boland, J. J.; Carver, P. H.; and Flynn, C. R. 1980. How much water supply capacity is enough. *Journal of the American Water Works Association* 72(7): 368–74.

California Department of Water Resources. 1985. *Emergency Water Shortage Manual* (draft). California Department of Water Resources.

Department of Commerce. 1973. *San Fernando, California Earthquake of February 9, 1971*. Vol. II, *Utilities, Transportation and Sociological Aspects*. U.S. Department of Commerce, NOAA. Washington: Government Printing Office.

Duke, M. C. 1981. An earthquake hazard plan for lifelines. In *Lifeline Earthquake Engineering, The Current State of Knowledge 1981. Proceedings of the ASCE Second Specialty Conference of the Technical Council on Lifeline Earthquake Engineering in Oakland, California, August 1981*, ed. D. J. Smith, Jr., pp. 1–16. New York: ASCE.

Engineering News Record. 1986. New seismic design plan redefines earthquake risk. *Engineering News Record* 216(5): 12.

Garland, S. B. 1980. Water rationing on Okinawa. In *Water Conservation Strategies*, pp. 6–7. Denver: AWWA.

Griffith, E. L. 1980. Southern California's Drought Response Program. In *Water Conservation Strategies*, pp. 39–43. Boulder: AWWA.

Hamilton, R. A. 1984. Thinking ahead, what will we do when the well runs dry? *Harvard Business Review* (November–December 1984): 28–40.

Harnett, J. S. 1980. Effects of the California drought on the East Bay Municipal Utility District. In *Water Conservation Strategies*, pp. 34–38. Boulder: AWWA.

Jahnus, R. H. 1979. Comments on social costs of future earthquakes. In *What Decision-Makers Need To Know: Policy and Social Science Research on Seismic Safety*, ed. S. Scott, pp. 2–5. Research Report 79-5. Berkeley: Institute of Governmental Studies, University of California.

Jezler, H. 1980. When the reservoir almost went dry. In *Water Conservation Strategies*, pp. 1–5. Denver: AWWA.

Miller, W. H. 1980. Mandatory water conservation and tap allocations in Denver, Colo. In *Water Conservation Strategies*, pp. 25–28. Boulder: AWWA.

Orange County Register. 1984. Water supply top concern in OC poll. *Orange County Register* (December 13, 1984): A24.

Patwardhan, A. S.; Cluff, L. S.; Power, M. S.; and Perman, R. C. 1981. Seismic risk reduction policy for lifelines. In *Lifeline Earthquake Engineering, The Current State of Knowledge 1981. Proceedings of the ASCE Second Specialty Conference of the Technical Council on Lifeline Earthquake Engineering in Oakland, California, August 1981*, ed. D. J. Smith, Jr., pp. 17–36. New York: ASCE.

Robie, R. B. 1980. California's program for dealing with the drought. In *Water Conservation Strategies*, pp. 29–33. Boulder: AWWA.

Rose, A. Z. 1981. Utility and economic activity in the context of earthquakes. In *Social and Economic Impact of Earthquakes on Utility Lifelines. Proceedings of the ASCE Construction Division Specialty Conference in San Francisco, California, May 1980*, ed. J. Isenberg, pp. 107–20. New York: ASCE.

Seih, K. E. 1971. A study of Holocene displacement history along the south-central reach of the San Andreas Fault. Ph.D. dissertation, Stanford University.

———. 1978. Prehistoric large earthquakes produced by slip on the San Andreas Fault at Pallett Creek, California. *Journal of Geophysical Research* 83(B8): 3907–39.

Theil, C. C. 1981. Lifelines, seismic hazards and public policy. In *Social and Economic Impact of Earthquakes on Utility Lifelines. Proceedings of the ASCE Construction Division Specialty Conference in San Francisco, California, May 1980*, ed. J. Isenberg, pp. 6–17. New York: ASCE.

Turner, R. H.; Nigg, J.; Paz, D.; and Young, B. 1979. Earthquake threat—

the human response in Southern California. Report prepared for the Institute for Social Science Research, University of California at Los Angeles.

U.S. Army Corps of Engineers. 1983a. *The State of the States in Water Supply/Conservation Planning and Management Programs*. U.S. Army Corps of Engineers, Engineer Institute for Water Resources, Policy Study 83-PS-1.

———. 1983b. *Evaluation of Drought Management Measures for Municipal and Industrial Water Supply*. U.S. Army Corps of Engineers, Engineer Institute for Water Resources, Contract Report 83-C-3. Report prepared by B. Dziegielewski, D. D. Baumann, and J. J. Boland.

———. 1983c. *Prototypal Application of a Drought Management Optimization Procedure to an Urban Water Supply System*. U.S. Army Corps of Engineers, Engineer Institute for Water Resources, Contract Report 83-C-4. Report prepared by B. Dziegielewski, D. D. Baumann, and J. J. Boland.

U.S. Water News. 1985a. Drought spurs some new ideas, new excuses. *U.S. Water News* 2(4): 3.

———. 1985b. N.Y. City gets tough with water-using lawbreakers. *U.S. Water News* 2(4):6.

Ward, D. B., and Taylor, C. E. 1981. Formulating Utah's seismic policy for lifelines (Utah Seismic Safety Advisory Council). In *Social and Economic Impact of Earthquakes on Utility Lifelines. Proceedings of the ASCE Construction Division Specialty Conference in San Francisco, California, May 1980*, ed. J. Isenberg, pp. 202–14. New York: ASCE.

Water Newsletter 1986. *Water Newsletter* 28(4).

Woo, V. 1982. Drought management: expecting the unexpected. *Journal of the American Water Works Association* 74(3): 126–31.

Wright, J. D. 1981. The state and local politics of seismic hazard. In *Social and Economic Impact of Earthquakes on Utility Lifelines. Proceedings of the ASCE Construction Division Specialty Conference in San Francisco, California, May 1980*, ed. J. Isenberg, pp. 193–201. New York: ASCE.

Young, G. K.; Taylor, R. S.; and Hanks, J. J. 1972. *A Methodology for Assessing Economic Risk of Water Supply Shortages*. U.S. Army Corps of Engineers, Engineer Institute for Water Resources.

# CHAPTER FIVE

# Social Objectives in Water Resources Planning and Management

Water plays a very significant role in the socioeconomic development of a country. In many cases, large water projects are often explicitly designed to manage natural systems so as to achieve major social objectives. Therefore, an increased awareness of social and environmental goals and objectives in water resources planning is essential to the success of a broad range of water projects all the way from small local projects to national programs.

For example, following the war in the Persian Gulf, there has been considerable discussion of how water might affect and enhance the prospects of the peace process in the Middle East. In Cyprus in 1986, the development of a common water supply through the efforts of the World Bank is credited with encouraging dialogue among different countries in the region. In the southeast African nation of Mozambique, a cease-fire after sixteen years of civil war coincided with the end of a drought that had ravaged the country for nearly a decade. When the recent rains transformed the parched countryside into abundant croplands, the war died faster than anyone had ever predicted.

Historically, in developing countries, the social goals of most water resources projects were to open unsettled portions of the country, to unify it and to enhance the national security. The driving social issue was the protection of public health. Once these goals had been accomplished, the economic development potential of water resources projects could then be fully realized. Production of hydro-electric power, agricultural production in reclamation areas, transportation savings through increased navigation, and the economic values inherent in flood control projects all increase in importance. Some concern may be expressed regarding social and environmental objectives (fish and wildlife protection, preservation of natural and scenic resources, provision of recreational amenities, etc.), but the primary emphasis was the economic recovery of the country. Water resources projects were seen as a means of stimulating employment, as an aid to economic recovery.

In industrial countries, however, water resources projects are also judged in terms of the substantial social and environmental costs that might have been considered before, but could not be quantified. The water supply planning process has therefore evolved from the focus on purely economic goals toward the multi-objective approach of attaining economic development and environmental protection goals.

## PREFERRED PLANNING APPROACH

Today, a broad approach to water supply planning is required. It has the objective not only of protecting public health, but also of protecting the waterways for their recreational amenities to include water-oriented recreation, body-contact water sports, and fishing. The improvement of water quality for the purpose of preserving the environment and of enhancing it is given equal weight to the historic goal of preserving public health.

Too often, water resources projects are governed by technical solutions to the exclusion of all other factors. Our present levels of technological understanding are already sufficient to allow us to present viable alternatives capable of getting at the tough water-related issues. However, in many cases, problems of water supply become defined more through a narrow understanding of possible technical solutions than through a broader understanding of social needs.

Today, more than ever, future water planners must be "society wise" as well as "technology wise." The benefits and costs associated with water resources development must include an analysis of the social and environmental impacts of all alternative solutions. Once alternative plans are formulated and their impacts to the public identified, feedback from impacted constituents must be obtained and incorporated into the planning process.

## SOCIAL AND ENVIRONMENTAL OBJECTIVES

To accomplish these objectives, the role of the civil engineer must be expanded to include the identification of social and environmental objectives. This stems from the recognition that civil engineers have a diversity of responsibilities regarding the objectives of water resources planning and management. First is the task of identifying societal objectives as a part of the planning process. Second is obtaining citizen reaction to proposed water projects. Lastly, although the engineer cannot unilaterally decree the objectives, he can appropriate means to make experience, skills, and ideas count in the societal process of formulating and altering such objectives.

Social and environmental objectives describe a variety of societal concerns which are usually not quantifiable in a monetary sense. Yet, it is necessary to address these objectives in making investment decisions regarding national, regional, and local infrastructure development. The objectives are identified by society as a whole rather than by separate professional or nonprofessional segments thereof; they evolve in time through a process of give and take among the various segments of society.

Society's basic goals change little with time. What does change are the objectives, policies, constraints and programs aimed at achieving these goals.

Table 1 contains a list of some recognized societal goals associated with water resources projects. Each goal has a corresponding social or environmental objective to address that goal and a program to meet that objective. Generally, a goal is addressed by a number of objectives (and may face several constraints, as well) and several programs might be required to satisfy completing each objective.

Objectives and constraints vary not only in time, but also in place, especially as cultural backgrounds vary from one country to another. They may also be influenced by the level of social organization—for example, local versus national objectives. One of the engineer's tasks when he proposes or undertakes a project is to be aware of all current societal objectives and constraints bearing on that project so that society's goals will be served. Although technology forms both the basis and a constraint as to what *can* be done, social perceptions and organization determine what *will* be done.

**Table 1. Societal Goals and Examples of Related Objectives and Water Resources Programs.**

| Goals | Social & Environmental Objectives | Programs |
|---|---|---|
| National Unity and Collective Security | Develop adequate communication. | Establish navigable waterways. |
| Economic Security | Reduce dependence on non-renewable resources. | Promote development of small hydropower facilities. |
| Freedom from Want | Encourage adequate food and fiber production. | Subsidize irrigation systems. |
| Public Health & Safety | Assure safe drinking water. | Regulate contaminants in input to and output from municipal water treatment plants. |
| Individual Freedom | Maximize opportunities for unrestricted fishing. | Protect natural backwater areas or create new ones at regional water projects. |
| Protect Ecological Systems | Reduce stream pollution. | Require permits for discharge into any water body under regional control. |
| Maintain Quality of Life | Provide recreational opportunities near urban areas. | Enable localities to establish recreational facilities at regional water projects. |
| National and Regional Prestige | Maintain unique regional characteristics. | Set aside lands to protect a scenic river from development. |
| Promotion of Humane Values Worldwide | Improve sanitary conditions in primitive places. | Train volunteers to plan small water and sewer systems using native materials. |

SOURCE: Adapted from American Society of Civil Engineers, 1984.

Some examples of how social and environmental objectives can be incorporated into regional programs include hydropower development, fish and wildlife preservation, recreational development of water projects, and programs for management of coastal zones.

Social and environmental objectives and constraints must be recognized at all phases of a project, but especially at the start of the planning process. The most effective means for recognizing objectives is by the use of an adequate public participation process. In such a process, the engineer and other knowledgeable professionals must *anticipate* the need to address objectives and constraints not apparent to the lay person. They must develop technically feasible alternatives which have a high probability of acceptance. Sooner or later, those objectives not addressed satisfactorily in planning, design, or implementation may return to adversely impact the project. The consequences can range from time delays to complete abandonment of the project.

Consider, for example, the following projects that have proceeded despite overwhelming evidence that social costs exceed social benefits:

- *Senegal River*, West Africa. Consortium of multilateral and bilateral lenders have spent $1 billion on two very large dams and related irrigation over the past 15 years. Riparian countries are now asking what to do with the water.
- *Gambia River*, West Africa. International river basin commission continues to push for large salt water barrage at Balingho in spite of studies showing economic infeasibility and likely ecological catastrophe.
- *Tana River*, Kenya, East Africa. Even though the catastrophic Bura Project nearly bankrupted Kenya, the Tana and Athi River Development Authority continues to push for the bigger Delta Project, with the urging of Japan.

In many cases, problems arise due to the appearance of incompatible objectives. For example, if a local community seeks to maintain a swimming beach in a reservoir, the potential for small hydropower development may be reduced. In another example, occupants of a floodplain think water should be detained upstream, but upstream residents are reluctant to provide the land needed for retention. In these situations, the engineer does not need to resolve the conflict—resolution is part of the decision process and responsibility lies with the political system.

When faced with incompatible objectives, what the engineer must do is to prepare alternate plans or scenarios, each accommodating conflicting objectives or constraints. The engineer must also help clarify the trade-offs associated with clearly articulated conflicts among alternatives. Even though the engineer may not make the final decision as to which, if any, alternative to pursue, he or she does have an obligation to advocate that alternate which seems to best fit what has been learned in the planning process. Ultimately, if the social and environmental objectives are incompatible with each other or with the purposes of the proposal, it may be necessary to point out that no plan is possible. If alternate plans are prepared, each alternate must be viable and not be presented merely to make the engineer's preferred plan appear more palatable.

Engineers are accustomed to beginning their efforts on a particular assignment by saying, "What is the engineering problem?" This is highly appropriate at some stage, but in the progression of steps required it ranks *behind* the establishment of goals, objectives and programs.

## THE NEXT TWENTY YEARS

In the next twenty years, the water industry will be caught up in a renewal of idealism with an emphasis on environmental quality and health. Demands for higher quality water will become universal. Regardless of local circumstances, the public will demand that everyone should have the best of the standards achieved elsewhere. Water quality will have to reach the same standard whether one is in Chicago, Cannes, Calcutta, or Canterbury. Water supplies will have to accomplish this objective in the era of instant global communication where the difficulties in one continent are known almost immediately in another.

This global rise in "consumerism" and "environmentalism" will lead to the emergence of risk assessment as one of the key issues in water supply. We have seen that in less developed countries the first priority has to be a regular supply of water free from the risk of waterborne disease. Concerns about the long-term ingestion of minute quantities of chemical and other contaminants are a lower order of priority. Even wealthy developed countries, however, will find the need to set priorities, to make realistic assessments of risk, and to objectively assess the cost-benefit of further purification measures. The final judgment about whether the benefits justify the risks will be a matter for each individual country to decide.

Another offshoot of the rise in environmentalism will be the creation of ecological engineering as a professional discipline equal to conventional fields of engineering such as civil, sanitary, environmental, etc. This discipline will evolve in response to the complexity of pollution problems faced by the professional engineering community. Ecological engineering combines environmental engineering with other disciplines such as biology, microbiology, chemistry, biochemistry, and ecology. Ecologically engineered systems will merge accumulated technical knowledge with natural science disciplines. Ecological engineering programs will become standard curricula, expanding from current offerings at Ohio State University, University of California at Davis, Oberlin College, and Massachusetts Institute of Technology (MIT).

Ecological engineering will find widespread applications in the wastewater treatment field. Ecologically engineered systems will use sound engineering in coordinating the interaction of appropriate ecosystems to degrade and remove waste materials from wastewater. Because such systems do not rely on harmful chemicals, there will be no risk of overdosing or undesirable residuals. Once established, ecologically engineered systems will be self-regulatory, reducing the opportunity for operator error. Tertiary levels of treatment and pathogen reductions will be accomplished within the treatment system.

## TECHNOLOGY-BASED INDUSTRY

Changes that society will experience in the next twenty years will be five times greater than that which we have witnessed over the last 100 years. The planet's exponential population growth will be matched, stride by stride, with an exponential growth in technology-based industry.

There will certainly be remnants of our operations and infrastructure—the investment we have made in our utility systems is simply too great to abandon no matter how improved the next generation of processes may be. But those remnants will be optimized to the maximum degree possible and will be integrated with a new generation of products.

Giant computers will control the distribution of finished and partially treated drinking water in dual distribution systems that cover river basins and, in some instances, entire states. New space-age piping material that will be molded in place and will exhibit all the attributes of stainless steel at one-half the weight and one-tenth the cost will gradually replace old water and sewer lines.

In the next twenty years, water utilities will depend on much more sophisticated laboratory support to identify problems and sustain quality treatment.

Corrosion control will be improved beyond the level of sophistication now often deemed warranted for simple protection of plumbing material as we move to minimize lead at the tap. Radon removal will require aeration by a majority of systems drawing water from the ground. Disinfection will have to be carefully controlled to minimize by-products while assuring public health protection for microbial contaminants. Ultrafiltration and nanofiltration for the removal of particles, colloids, bacteria, viruses, and organic molecules will be universally practiced throughout the water supply industry.

As we move into the twenty-first century, old technologies will be applied in new and more sophisticated ways to public water supplies. An expert system will optimize the membrane pretreatment processes which will occur in the fifty-year-old rapid sand filtration plant and fifteen-year-old granular activated carbon absorbers.

Small systems will find that highly engineered solutions that simply downsize approaches used by large systems will cost too much per capita. Package technologies installed by vendors will be far less costly. They will also find innovative approaches to financing, operation, and maintenance that often depend on contracting with outside parties. They will return to greater use of slow sand filtration for its simplicity, perhaps with standardized designs. Highly concentrated sunlight will be used to purify contaminated water in arid areas where sunlight is abundant. Such photo degradation of polluted waters will find widespread applications in rural areas in the Middle East where more advanced technologies are impractical.

Water shortages, particularly in areas of high growth and low rainfall, will drive use of comprehensive water conservation measures. These will range from low-flush toilets and sprinkler systems controlled by moisture-sensing devices to simple efforts to encourage shorter showers and plant lawns suitable to a dry environment. Utilities will also move to more dual piping systems providing high quality water for use inside homes, and lower quality water for use outside homes. More and more membrane systems will treat water with high dissolved solids when quality sources are not available.

With currently available technologies, agricultural water use can be cut 10 to 50 percent, industrial use 40 to 90 percent, and municipal use by one-third with no sacrifice to economic output or quality of life. A study by the East Bay Municipal Water District in Oakland, California, indicates that homes with water conserving landscapes use up to 42 percent (or 209 gallons a day) less water than those with traditional landscapes.

The next twenty years will see water-conserving landscaping becoming mandatory for all consumers. Water-intensive traditional landscaping will be reserved for the very rich (who could afford the steep surcharges for excessive water use) or limited to public facilities where they can be enjoyed by the general public.

Water recycling will play an increasing role in municipal water supply. Much of the world will copy the Japanese who have become adept at recycling. In Japan, basements in apartment buildings are already equipped with minipurification systems that microfilter wastewater and then recirculate the stream for sanitary purposes after it has been chlorinated. In Tokyo, the law states that all apartment blocks of more than fifteen floors must be equipped with recycling equipment. In Tokyo's business district of Shinjuku, a group of about two dozen buildings direct their wastewater to a centralized minipurification station that is cooperatively owned.

## LONG-TERM SUPPLEMENTAL WATER SUPPLIES

The world's burgeoning population will spawn the development of long-term supplemental water supplies to meet increasing demands for water.

We can expect to see artificial islands constructed off-shore of major population centers. These man-made islands will contain complete water and power process systems: an entire industrial complex that would handle society's basic needs for food, water and power. A nuclear breeder-reactor will generate power for a unit that desalinates salt water which will be linked to a mineral recovery system that captures the valuable mineral by-products of the desalination. The waste heat would be used for industrial uses like hot houses for growing food. Excess energy could be pumped ashore to heat and cool cities before returning to the island as a coolant for the breeder-reactor.

For less-developed countries, however, reducing their rate of population growth will be the only means of managing already strained water supplies.

In the next twenty years, some water-short locations will be practicing direct reuse of municipal wastewater, most likely following reverse osmosis treatment. Innovative water management practices will likely occur in the following geographical areas:

Along the western coast of the United States, supertankers will carry 250,000 acre-feet per year of freshwater from Juneau, Alaska, to thirsty communities along the California coast. At $1,500 to $2,000 per acre-foot, this water will be cheaper than desalination and diversion projects, will be of superior quality, and will have no capital investment costs. Other proposals, such as an Alaska-California pipeline, have been estimated to cost upwards of $150 billion and would have resulted in significant environmental impacts.

The Alaskan freshwater will be collected from sources with minimal environmental impact: coastal lakes discharging almost directly into the ocean with no anadromous fish runs due to the height of the discharge. Area-of-origin impacts will be virtually nonexistent: Alaska discharges over one billion acre-feet of water per year, half of which is in the southeast where rainfall ranges to 400 inches per year.

In the United States, the water supply industry, not the federal government, will be in the leadership role of developing new regulations and standards. This self-regulation will be a natural progression of the industry's quest to understand everything about its product and continually improve its service to the consumers. Consumers will be paying $5 per cubic meter of water (yes, the United States will have finally switched to the metric system) and surveys will indicate their willingness to pay twice that amount for an assured supply of drinking water.

Laws and regulations that promote the development of a viable water market will be enacted, allowing water to be traded as a commodity among users. This will permit the re-allocation of water to new, more productive uses in response to socioeconomic changes.

Japan and the United States will continue to increase their industrial output while at the same time reducing the amount of water consumed for industrial processes. Water pollution control laws will provide the main impetus for increase industrial water recycling in most of the world's wealthier countries.

Extensive recycling will allow Japan to meet its goal of serving 99 percent of its population by a public drinking water supply. The cost of drinking water will rise dramatically. A liter of drinking water in Tokyo will cost about $5, or more than five times the price of bottled water in Paris.

In Europe, the complexity of water management problems and river systems within the continent will lead to the use of a single computer model to suit every individual system application for all river basins. This unified approach will lead to environmentally sound management of water resources, including groundwater, lakes, reservoirs, rivers, and estuaries. Customized software systems will allow users to solve a variety of problems associated with water quantity or quality. These systems will be designed to integrate geographic information systems (GIS), large databases, simulation models, and satellite imagery.

In Central America, simple, low-cost, and environmentally sound concepts such as gray water filters, composting toilets and wetland treatment will be installed in rural communities to solve local sanitation problems.

Due to its warm climate and abundant labor force, Mexico will emerge as a world leader in the export of fruits and vegetables. Two decades of foreign investment in irrigation technology and water-efficiency farming practices will result in record harvests which will be in great demand by Japan and Eastern Europe. Mexican agricultural exports will be marketed worldwide.

Mexico will lead the world in the so-called "Green Revolution," the development of new crop varieties more tolerant to drought. New varieties of corn developed at the International Maize and Wheat Improvement Center in El Batan, Mexico, are already capable of providing at least 30 percent more grain under midseason drought than farmers would obtain using nontolerant varieties. Such crops not only hold promise for dry areas of the globe, but will also receive widespread application in traditional growing areas such as the Corn Belt in the United States.

In the next twenty years, Mexico will emerge as one of the most active construction markets in the world. According to U.S. Department of Commerce estimates, Mexico's imports of construction machinery and equipment will increase a staggering 24.9 percent a year during 1992–1994, from a mark of $453.1 million in 1991. Construction activity has ballooned from a contract volume of $2 billion in 1986 to $18 billion in 1991.

With the North American Free Trade Agreement effective on January 1, 1994, the world's largest trading block will be created, with a combined population of 362 million and combined gross domestic product (GDP) of around $6 trillion. Mexico has almost a quarter of that population, but only a twentieth of the GDP. The disparities in infrastructure are striking, with nearly 40 percent of Mexico's people living in poverty, amid chronic shortages in housing, water and sewage systems, sanitation services, environmental protection and electrical energy. As part of the Mexican government's strategy of encouraging private sector participation in public infrastructure projects, a majority of the country's regional water systems will be offered on a concession basis.

As a result of these varied practices, in every part of the world each consumer will have to pay for the true value of the water commodity. Water prices will escalate around the globe. Wasteful water practices will have to be curtailed. Existing water systems will be rebuilt to detect and eliminate leaks and optimize system operation.

Finally, our continued production/pollution cycle will lead to the concept of "manufactured" drinking and cooking water in the twenty-first century. For each source of water, an industrial process would remove everything, and then introduce desired elements and compounds back into the water. Thus, we might expect a new sophisticated drinking water manufacturing technology which would market waters with special medical, nutritional, and safety characteristics.

## REFERENCES

*Measuring the Benefits of Clean Air and Water, Resources for the Future, Inc.* Washington DC, 1984.

*The Role of Social and Behavioral Sciences in Water Resources Planning and Management.* Water Resources Planning and Management Division, American Society of Civil Engineers, New York, 1988.

*Social and Environmental Objectives in Water Resources Planning and Management, Committee on Social and Environmental Objectives.* American Society of Civil Engineers, New York, 1984.

*Social Feasibility as an Alternative Approach to Water Resources Planning.*

Virginia Water Resources Research Center, Virginia Polytechnic Institute and State University, Blacksburg, Virginia, 1985.

Landscaping Can Cut Use by 42%. *U.S. Water News*, April 1993, Volume 9, No. 10, Page 9.

Manwaring, James F. Metamorphosis, *Water Engineering and Management*, July 1990, Page 14.

New Water Shortage Strategies, Lewis, William. *Water Engineering and Management*, July 1990, Page 15.

Cook, Michael. New Use of Old Technologies for Safe Drinking Water. *Water Engineering and Management*, July 1990, Page 18.

Israeli Researchers Use Sunlight to Remove Pollutants. *U.S. Water News*, March 1993, Volume 9, No. 9, Page 2.

California Firm Files for Rights to Allow Transfer from Alaska. *U.S. Water News*, April 1992, Page 10.

New, Disturbing Trends Emerging in Evolution of Water Management. *U.S. Water News*, February 1993, Volume 9, No. 8.

Water Recycling Is a Way of Life in Japan. *U.S. Water News*, February 1993, Vol. 9, No. 8, pg. 25.

Flexible Software Development for European Water Management. *U.S. Water News*, April 1993, Vol. 9, No. 10.

Organization Promotes Simple Third World Water Solutions. *U.S. Water News*, April 1993, Vol. 9, No 10.

Peace Follows Rain in Mozambique. *U.S. Water News*, May 1993, Vol. 9, No. 11, pg. 2.

Carney, Michael. European Drinking Water Standards. *Journal American Water Works Association*, June 1991, pg. 48–55.

Postel, Sandra. Last Oasis. Facing Water Scarcity. Worldwatch Institute.

Will Water Supplies Go the Way of the Buffalo? *U.S. Water News*, May 1993, Vol. 9, No. 11, pg. 12.

# BIBLIOGRAPHY

This bibliography includes papers, reports, and policy studies published by the U.S. Army Corps of Engineers' Engineer Institute for Water Resources. These works are a valuable source of information for the water planner.

1972. *A Methodology for Assessing Economic Risk of Water Supply Shortages.* U.S. Army Corps of Engineers, Engineer Institute for Water Resources, Contract Report 72-6. Report prepared by G. K. Young, R. S. Taylor, and J. J. Hanks.

1973. *Evaluation of Quality Parameters in Water Resources Planning—a State-of-the-Art Survey of the Economics of Water Quality.* U.S. Army Corps of Engineers, Engineer Institute for Water Resources. Report prepared by E. D. Bovet.

1973. *A Study of How Water Quality Factors Can Be Incorporated into Water Supply Analysis.* U.S. Army Corps of Engineers, Engineer Institute for Water Resources, Contract Report 74-2. Report prepared by Ernst and Ernst.

1975. *Economic Concepts and Techniques Pertaining to Water Supply, Water Allocation, and Water Quality.* U.S. Army Corps of Engineers, Engineer Institute for Water Resources, Paper 75–P-5. Report prepared by E. D. Bovet.

1979. *The Role of Conservation in Water Supply Planning.* U.S. Army Corps of Engineers, Engineer Institute for Water Resources, Contract Report 78-2. Report prepared by D. D. Baumann, J. J. Boland, J. H. Sims, B. Kranzer, and P. H. Carver.

1980. *The Evaluation of Water Conservation for Municipal and Industrial Water Supply: Procedures Manual.* U.S. Army Corps of Engineers, Engineer Institute for Water Resources, Contract Report 80-1. Report prepared by D. D. Baumann, J. J. Boland, and J. H. Sims.

1981. *An Assessment of Municipal and Industrial Water Use Forecasting Approaches.* U.S. Army Corps of Engineers, Engineer Institute for Water

Resources, Contract Report 81-C-5. Report prepared by J. J. Boland, D. D. Baumann, and B. Dziegielewski.

1981. *Selected Works in Water Supply, Water Conservation and Water Quality Planning*, ed. J. E. Crews and J. Tang. U.S. Army Corps of Engineers, Engineer Institute for Water Resources, Research Report 81-R-10.

1983. *The State of the States in Water Supply/Conservation Planning and Management Programs*. U.S. Army Corps of Engineers, Engineer Institute for Water Resources, Policy Study 83-PS-1.

1983. *Forecasting Municipal and Industrial Water Use: A Handbook of Methods*. U.S. Army Corps of Engineers, Engineer Institute for Water Resources, Contract Report 83-C-1. Report prepared by J. J. Boland, W. S. Moy, J. L. Pacey, and R. C. Steiner.

1983. *Evaluation of Drought Management Measures for Municipal and Industrial Water Supply*. U.S. Army Corps of Engineers, Engineer Institute for Water Resources, Contract Report 83-C-3. Report prepared by B. Dziegielewski, D. D. Baumann, and J. J. Boland.

1983. *Prototypal Application of a Drought Management Optimization Procedure to an Urban Water Supply System*. U.S. Army Corps of Engineers, Engineer Institute for Water Resources, Contract Report 83-C-4. Report prepared by B. Dziegielewski, D. D. Baumann, and J. J. Boland.

1984. *Influence of Price and Rate Structures on Municipal and Industrial Water Use*. U.S. Army Corps of Engineers, Engineer Institute for Water Resources, Contract Report 84-C-2. Report prepared by J. J. Boland, B. Dziegielewski, D. D. Baumann, and E. M. Opitz.

1985. *1985 Annual Conference Proceedings, Managing a Priceless Resource, Washington, D.C., June 23–27, 1985*. American Water Works Association, Denver, CO.

1987. *Resolving Locational Conflict*. Center for Urban Policy Research; Rutgers—The State University of New Jersey.

1988. *The Water Planet*. Watson, Lyall, Crown Publishers; New York, NY.

1988. *The Role of Social and Behavioral Sciences in Water Resources Planning and Management, Proceedings of an Engineering Foundation Conference in conjunction with the Universities Council on Water Resources; Santa Barbara, California, May 3–8, 1987*. American Society of Civil Engineers; New York, NY.

1989. *Balancing the Needs of Water Use*; Moore, James W., Springer-Verlag; New York, NY.

1989. *Water Resources Planning and Management, Proceedings of the 16th Annual Conference, May 21–25, 1989*. Water Resources Planning and Management Division, American Society of Civil Engineers. New York, NY.

1990. *The Water Encyclopedia*. Leeden, F., et al. Lewis Publishers; Chelsea, Michigan.

1990. State of the World 1990, Worldwatch Institute. W. W. Norton & Company, Inc.; New York, NY.

1991. State of the World 1991, Worldwatch Institute. W. W. Norton & Company, Inc.; New York, NY.

1992. State of the World 1992, Worldwatch Institute. W. W. Norton & Company, Inc.; New York, NY.

1993. *Water Management in the '90's, A Time for Innovation, Proceedings of the 20th Anniversary Conference, Seattle, WA, May 1–5, 1993.* Water Resources Planning and Management Division, American Society of Civil Engineers; New York, NY.

# Index

Page numbers in *italic* indicate illustrations or tables.

Accceptable risk, estimating, 208–210
Accounting systems, and water pricing policy, 157, 161, 162
Afghanistan, 5
Kabul, 5
Africa, 2
Aggregate demand curve for water, 174, *175*
Aharoni, Y., 206
Air conditioning, 40
Alaska, 248
Allison, R. C., 193, 206
Alternative futures, 89
Amazon River, 3
American Water Works Association, 60, 61, 65, 66, 67, 113, 193, 226
Andes, 3
Anton, W. F., 222
Apartments and condominiums, 37
ARIMA (Auto Regressive Integrated Moving Average) model, 95–96
Army Corps of Engineers, U.S., 127, 217, 227
Asian Development Bank, 5

Australia, 16–17
Darling River, 16
Autocorrelation, 99

Baumann, D. D., 136
Bettinger, R. W., 220
Billing system, data collection with, 80, 149
Blackburn, A. M., 225
Blocks of water, and varying rates, 165–67
Boland, J. J., 25, 75, 90, 179, 185, 208
Bovet, E. D., 126, 128, 131–132
Bovet Water Quality Index, 131–132
Box-Jenkins technique, 95
Brazil, 224
Brewer, R., 40, 43, 44, 47, 48–49

Calculated risk, identifying, 207–208
California,
drought in, 225–226
earthquake losses in, 213–220
historical earthquakes in, *214–215*

urban water production and
losses in, *61*
California Department of Water
Resources, 192
Office of Water Conservation,
60, 61, 64, 68
Carver, P. H., 208
Central America, 249
Chemical industry, water use in,
46
China, 2, 3, 13–14
Yangtze River, 3, 13, 14
Coelen, S. P., 29
Collins, M. A., 90
Colorado, 226
Commercial water demands/use,
49–58
commercial SIC classifications,
*51*
forecasting applications, 56–58
meter errors and, 62
summary of diurnal peaking, *25*
types of, 50–56
water conservation in, 119–121
water use in, *52–54, 120*
Common lot areas, 37
Components of water demand, 35–
69
commercial demands, 49–58
industrial demands, 38–49
residential demands, 35–38
unaccounted-for water use, 59–
69
Computer(s),
control of irrigation, 118
failure of, as a threat to water
lifelines, 198
Connecticut, regionalized water
supply in, 31
Conservation. *See* Water conservation

Consolidation of water utilities,
31–32
Contaminant, water use classification by, 126–127
Contingency tree methods of forecasting, 89–90, 92, 93
Cooling water, 40, 44
Corpus Christi, Texas, drought in,
226
Corrosion control, 246
Craddock, E., 164
Crews, J. E., 111
Criteria for water demand, 19–34
forecasting applications, 25–27
regionalization, 27–32
spatial and temporal variations,
19–26
water-supply planning philosophies, 32–34
Cyprus, 239

Daily peaking of water demand, 21
Danielson, L. E., 86–87
Database files,
on place/time/class of water use,
80–82
Data collection on forecasting, 75–
83
contingency planning for
droughts and, 231–35
data collection efforts, 78–80
data types and sources, 76–79,
*233*
defining the study area, 75–76
organizing local-level data, 80–
83
Seattle case study, 142–51
Data disaggregation, 79–80, 148
DeKay, F. C., *88*, 142
Denver, Colorado, 226
Desalination projects, 138

Disasters, natural. *See* Hazards to
water lifeline
Distribution system, water, leaks
and breakage in, 64–69
Diurnal peaking of water demand,
23–26
Dowdy, S., 97
Drought, 224–225
conservation carry-over after,
33–34
criteria and indicators, 226–228
data requirements for managing,
231–35
historical background on management of, 224–26
perspectives on contingency
planning for, 229–31, 261
traditional approach to management of, 228–29
Duke, M. C., 208, 218
Dworkin, D. M., 139–41

Earthquake threat to water lifelines, 194–95
case studies on reducing, 220–
24
energy relationships in, *195*
historical background on, 212–
20
intensity scale and Richter scale
relationships, *216*
Mercalli intensity scale, *217–18*
risk reduction planning, *211*
risk zones in the U.S., *194*
East Africa, 243
Tana River, 243
East Bay Municipal Utility District
(EBMUD) seismic hazard mitigation case study, 222–24
water conserving landscapes
study, 247

Ecological Engineering, 244–245
Econometric forecasting, 97–106
advantages/disadvantages, *109*
step-by-step, 100–106
El Salvador, 212–213
End-use forecasting, 94–95, 109
Energy production, water consump-
tion for, 38, 39
Environmental Protection Agency,
133, 134
Exponential demand curve for
water, *180*

Federal Emergency Management
Agency, 195, 203
Financing water supply, *See also*
Water industry economics,
and price elasticity
capital improvements and, 28
impact of hazard mitigation on,
202–04
Floods, 196
Flynn, C. R., 208
Food processing, water use in, 46
Forecasting water needs, 72–152
commercial use applications,
56–58
conclusions on, 148–51
data collection on, 65–73
flowchart for, *74*
impact of water conservation on,
110–125
impact of water quality on, 125–
41
industrial use applications, 47–
49
integrated approach to, 141–48
methods for, 83–93
residential use applications, 38
role of pricing policies in, *154*
techniques of, 94–110

water-demand criteria and, 25–27
and water supply availability in droughts, *234*
France, 11–12
    Compagnie Generale des Eaux, 11
    Lyonnaise des Eaux-Dumex, 11
    Ministry of Health, 12
Fredette, J. B., 61

Garland, S. B., 225
George, S. S., 94, 95, 96, 98–100, 106–107, 108
Germany, 10–11
Geyer, J. C., 23, 49
Glass/glasswear industry, water use in, 47
Goldstein, J., 153–58, 161, 163, 164, 170
Great Britain, drought in, 225
Griffith, E. L., 226

Hamilton, R. A., 226
Hanke, S. H., 187
Hanks, J. J., 208
Hanson, H., 28, 31
Harnett, J. S., 226
Hazardous materials, spills/dumping, 197–98, 199
Hazards to water lifeline, 193–02
    disaster effects matrix, *201*
    frequency of natural disasters, *203*
    interrelationships of natural disasters and effects, *200*
    and public policy, 204–211
Heteroskedasticity, 99
Hildebrand, C. E., 186, 191
Honolulu Board of Water Supply, 81–83

Horticultural practices, and water use, 116–118
Housing and Urban Development, U.S. Department of, 112, 114
Howe, W. J., 241, 244, 245–48
Hrezo, M. S., 241, 244, 245–48
Hudson, W. D., 63
Hurricanes, 196, 216
Hybrid models of forecasting, 106–107, 109

Income, increasing, and water demand, 174, 176
India, 2, 4–5
    National Commission on Floods, 4
Indoor vs. outdoor residential water use, 35, 36, 37–38, *117*
Industrial water demands/use, 38–49
    factors affecting, 44–47
    forecast guidelines, 47–49
    manufacturing classifications and, *41*, 42
    meter errors and, 62
    per employee by industry, *48*
    summary of water use, *52–54*
    types of, 40
    water conservation in, 119
    water reclamation in, 136–39
    water recycling, 42–44, *45*, 119
Institutional structures, and water pricing policies, 156, 158
Institutional water, use, summary of diurnal peaking, *25*
Intensity-of-use, as a price variable, 172
Inter-American Development Bank, 16
Interviews, in social-acceptability studies, 255, 256–57
Iran, 6–7

Irrigation, 37, 115
  irrigated acres in the U.S., *39*

Japan, 247–249
Jezler, H., 224
Johnstone, G. W., 28, 30, 31–32
Jones, C. V., 148

Kenya, 243
Kern County, California, earth-
  quakes, 219
Kim, J. R., 49–58, 119
Koller, K. L., 40, 43, 44, 47, 48–
  49

Lake Huron, 2
Leakage and breakage, water loss
  by, 64–69
  detection of, 67–69, 119
Lifelines, 192. *See also* Water-lifeline hazards mitigation
Linaweaver, F. P., 23, 49
Linear demand curves for water,
  *179, 180*
Long Beach, California earthquake, 219
Low-flow water fixtures and appliances, 202–205

McCuen, R. H., 49–58, 119
Malawi, 3
Malloy, C. W., 90
Marginal-cost pricing of water,
  168–169
Marin (California) Municipal Water
  District, 33, *34*
Market-demand curve, 174, 175
Mercalli intensity scale, *217–218*
Meters,
  errors in, and water loss, 62–64
  pricing policies and, 157, 162

underregistration of meters, 63,
  *64*
Mexico, 15–16, 212, 249–250
  Mexico City, 15
  Cutzamala Valley, 15
  Coatzacoalcos River, 15
  Pennex, 16
  Teapa River, 16
  Gopalapa River, 16
Micro time-series methods of fore-
  casting, 86–89, *92*, 93
Middle East, 2, 3, 239, 246
Miller, W. H., 226
Mitre Corporation, 132–34
Moyer, E. E., 60, 65, 66, 67
Morocco, 3
Mozambique, 239
Mugler, M. W., 111
Multicolinearity, 99, 100
Multiple-coefficient methods of
  forecasting, 86, 92, 93

National Sanitation Foundation,
  135–136
National Water Commission, 42
National Weather Service, 107
Need, as a price variable, 172
New Jersey, regionalized water
  supply in, 31
New York, 170
Nilson, D. C., Jr., 244–45
Normalization forecasting, 107–
  108
North American Free Trade Agree-
  ment (NAFTA), 250
Nuclear power plants, water use
  by, 38, *39*

Okinawa, drought in, 225
Orange County Municipal Water
  District, 164

Pacific Gas and Electric Company, case study on seismic hazard mitigation, 220–221
Pacific Northwest River Basin study, 145
Palmer Index and drought, *227*
Paper products industry, water use in, 46
Patwardham, A. S., 207
PDI (Prevalence-Duration-Intensity) water quality index, 132–134
Peak-flow observations, daily/seasonal, 21, *22* diurnal, 23–26
Peak-load pricing, 167–68
Per capita methods of forecasting, 84
Per connection methods of forecasting, 85
Petroleum refining, water use in, 46–47
Philadelphia, unaccounted-for water loss in, 60–61
Philippines, 7–8
   National Water Resource Council, 7
   Manila Bay, 8
   Department of Environmental and Natural Resources, 8
Philosophy of water-supply planning, 32–34
Plummer, A. H., Jr., 90
Poland, 12–13
   Vistula River, 12, 13
Policies. *See* Public policies
Political feasibility model of public participation, 248, *249*
Politics of water, pricing policies and, 157–58
Pollution Index (PI) of water quality, 134–35
Price elasticity of demand, 170–91
   definition of, 173–79
   factors influencing, 179–84

method for determining, 184–86
role of price, 172–73
sectoral water use and, *186*
water industry economics and, 186–191
Price level, 163–64
Price structure, and water rates, 163–70
   common, *163*
   fixed charges, 164
   flat or uniform rate, 164–65
   marginal-cost pricing, 168–69
   peak-load pricing, 167–68
   summary of, 170
   varying rates, 165–67
Primary metals, water use in, 47
Process water, 40
Public policies on hazard mitigation, 204–211
   risk-reduction programs, 207–211
   wastewater systems, 206
   water lifelines, 206
Public policies on pricing water, 153–63
   history of, 153–55
   insstitutional structures responsible for, 156
   obstacles to setting adequate, 156–57
   procedure for implementing full-cost, 161–63
   subsidized v. self-sustaining, 158–61
Public Water System Coordination Act of 1977, 31

Rates for water. *See* Water rates
Recycling water, in industry, 42–44, *45*, 119
Regionalization of water-supply facilities, future trends in, 31–32

problems faced by small water systems, 28–29
requirements for effective, 30–31
types of, 29–30
Regression analysis. *See* Econometric forecasting
Residential water demands/use, 35–38
conservation methods, 112–119
forecast applications, 38
meter errors and, 62, *63*
per capita interior use for conserving vs. nonconserving homes, *36*, 37
typical hourly, *24*
Residual risk, estimating, 208–210
Richter scale, *216*
Riots/vandalism/civil disorder/work stoppages,
as a threat to water lifeline, 197
Risk management in water-supply systems,
pressure to establish "no-risk," 205–206
risk reduction programs, 207–211
risk reduction through regionalization, 29–30
seismic risk reduction, *211*, 212–214
Robie, R. B., 226
Rose, A. Z., 202
Russia, 14–15
Lake Baikal, 14, 15
Selinga River, 15
Russell, J. D., 167–68

Saleba, G. S., 73, 74, 76, 77, 79, 100, 107, 108
San Francisco, California,
drought in, 226
earthquakes in, 218, 219

seismic hazard mitigation case study, 222–224
Season of the year,
peaking of water demand, 21, *22*
and price elasticity of demand, 181–182
Seattle Water Department case study, as an integrated approach to water demand forecasting, 142–48
comprehensive plan, 147–48
conclusions on, 148–51
earlier forecasting models, 142–45
evolution of techniques/databases, *143*
metropolitan water-supply study, 146
rate study, 146–47
River Basin Study, 145
Seih, K. E., 212
Shelstad, M. J., 28, 31
Shields, J., 193, 206
Shopping centers/malls, water use in, 50–51, 55–56, *57*
Single-coefficient forecasting methods, 84–85, 91, 92, 93
Social alternatives, in water conservation, 118–119
Social objectives in Water Resource planning and management, 239–244
preferral planning approach, 240
social and environmental objectives, 241–244
South Africa, 3
Spatial variations in water demand, 19–26
Standard Industrial Classification System (SIC), *41*, 42
for commercial businesses, 50, *51*

Steinbeck, John, v
Sudan, 3
Sutherland, R. C., 49–58
Syracuse University Civil Engi-
    neering Department's water
    quality index, 134–35

Taylor, C. E., 221
Taylor, R. S., 208
Temporal variations in water de-
    mand, 19–26
Textiles industry, water use in, 46
Thiel, C. C., 203
Time-series analysis. *See* Micro
    time-series methods of fore-
    casting
Time-series forecasting technique,
    95–97, *109*
Tornadoes, 196
Toxic materials, 197, 198, *199*
Tsunamis, 196–207
Turner, R. H., 205

Unaccounted-for water use (UAW),
    59–69
    authorized vs. unauthorized, 59–
    61
    types of unauthorized, 61–69
U.S. Department of Commerce,
    249
United States, earthquake loss in,
    213–220
United Nations, 2
Urban water supply. *See* Water-
    supply systems
Users of water. *See also* Water
    consumption; Water use
    class of, and price elasticity,
    179–81
    response of, and price elasticity,
    182–83
Utah, seismic hazard mitigation
    case study in, 221–222

Viet Nam, 8–10
    Mekong River, 9
    Hong River, 9
    Mekong Delta, 9

Ward, D. B., 221
Washington, regionalized water
    supply in, 31
Wastewater systems, 206
Water, unlike other goods/com-
    modities, 187–88
Water audit, 66
Water conservation,
    carry-over effect after drought,
    33–34
    defined, 110–111
    effect of, on demand curve, 174,
    *176*
    forecasting applications, 121–25
    impact on water-demand forcast-
    ing, 110–125
    incentives provided by rate
    schedules, *171*
    increasing-block rates and, 166–
    67
    measures of, 111–121
    per capita consumption in con-
    serving/nonconserving resi-
    dences, *36, 37*
Water consumption. *See also* Water
    use
    in commercial businesses, *52–
    54, 120*
    in conserving vs. non-conserving
    homes, *36, 37, 117*
    of employees by industry, *48*
    factors affecting, per capita, 20–
    24
    in industry, *52–54*
    place of use, 81–82
    time of use, 82
    unaccounted-for. *See* Unac-
    counted-for water use (UAW)

Water demand, 19–71. *See also* Forecasting water needs
  alternatives in, and water reclamation, 138–39
  components of, 35–69
  criteria for, 19–34
  demand curve slopes, *177, 179, 180*
  effect of, on water price, 189
  factor influencing urban, *77*
  price elasticity of, 173–79
  uncertainties in, 190–91
Water industry economics, and price elasticity, 186–91
Water-lifeline hazards mitigation, 192–238
  drought management, 224–35
  economic impact of, 202–04
  lifeline system reliability levels, *210*
  public policy and hazards, 204–211
  seismic hazard mitigation case studies, 212–24
  survey of hazards, 193–202
Water pressure, reductions in, 115–116
Water pricing, 153–91. *See also* Water rates
  price elasticity of demand, 170–91
  pricing policies, 153–63
  rates and price structures, 163–70
Water quality,
  and classes of water use, 125–29
  forecasting applications, 139–41
  impact of water reclamation on, 136–39
  indices of, 129–136
Water Quality Index (WQI), 135–36
Water rates. *See also* Water pricing

artificially low, 28–29
  fixed charge, 164
  flat or uniform, 164–65
  improving the rate-making process, 162
  and price structures, 163–70
  structure characteristics of, and price elasticity of demand, 184
  study of, in Seattle, 146–47
  varying (declining-block/increasing-block), 165–67
Water reclamation, impact on water quality, 136–39
  forecasting applications, 139–41
Water supply,
  alternatives for balancing supply and demand of, and water reclamation, 137, *138*
  illustrative list of emergency, *235*
  methods for forecasting availability of, *234*
  steps for predicting deficits in, *209*
  uncertainties in demand and, 189–91
Water-supply systems. *See also* Water-lifeline hazards mitigation
  cost of physical facilities, 188
  philosophy of planning, 32–34
  regionalization of, 27–32
  risk management in. *See* Risk management in water-supply systems
  stresses to, ix–x
Water treatment,
  classification by cost of, 128–29
  commonly used, *130*
Water use. *See also* Users of water; Water consumption
  classes of, 81, 125–29

sectoral, and price elasticity
    range, 186
Wearden, S., 97
Weather,
    drought indicators and, 226–28
    modification, 138
    and water-demand forecasting,
    107
West Africa, 243
    Senegal River, 243
    Gambia River, 243
Whitford, P. W., 90
Wolff, J. B., 23, 49

Wolman, A., 187
Woo, V., 34, 226, 229, 231
World Bank, 5, 16, 239
World Resource Institute, 2
Worldwatch Institute, 3
Wright, J. D., 205

Xeriscapes, 118

UNDP, 4
UNICEF, 4

Young, G. K., 208